Prologue

The story is about the first concert performed as part of the 'Grand gala opening' of a newly and uniquely built, futuristic Amphitheater. The concert, with all the delectable musical repertoire, is interspersed with several episodes of drama, mystery, surprises, and suspense. Spectacular 3D Laser light shows, Television, audiovisual, and multimedia effects embellished the settings and enhanced the prevailing ambiance. Several odd events sporadically popped up along the way. Gradually, they evolved into rather significant parts of the event itself. Rather strangely, its aftereffects linger on further in tantalizing ways. The obsessions with the true origin and secrets of the mysteries created during the concert nag many curious minds long after the conclusion of the concert. It assumes all the trappings of a folk story that does not seem to have an ending.

The splendid concert hall, unlike any others, provides the audience a fully '*immersive*' (able to be an integral part of it) experience. They are entertained, pampered, and enamored to unprecedented levels; ultimately making them active participants in the ongoing performance and drama. The concert hall, itself is a "*Xanadu*" (a place of great marvels and contentment) of *par excellence* (a degree of excellence, beyond comparison) in terms of its architecture with ornamental facades and colonnades reminiscent of Indian & Mughal architecture. The many fluted and glazed terracotta columns and vaulted ceilings display inscribed Vedic verses, arts, and symbols adorning them. The wall tapestries depict '*Mohiniyattam*' & '*Bharatha-Natyam*' dance steps & poses on one side, and paintings of pompous processions of maharajahs carried on palanquin; the splendor of palaces in juxtaposition with struggles of the poor subjects; parades of the clowns & artisans in folk festivals on the opposite sidewalls. The architectural fittings and seatings are custom fabricated to aid and contribute to the overall superb acoustic qualities with '*beyond the state-of-the art*' technical perfection. The hall and surroundings are well-appointed (abundantly supplied) with futuristic 3D visual and '*surround sound*' listening features. They are thematically and systematically integrated with advanced light, visual, and television production techniques & sound effects. These '*eclectic*' (employing individual elements from a variety of sources, systems, & styles) components help to intensify listeners' musical acumen and appreciation of every aspect of the performance. It elevates the psyche and imagination of everyone; evokes emotional effect to the highest degree attainable!

The visual and auditory circuits reside in our brains in proximity to each other. Crosstalk, mutual stimulation, and interactions occur naturally. The features designed into the auditorium tend to accentuate (intensify, emphasize) these faculties to unprecedented levels to afford the utmost in listening pleasures and visual stimulation. In order to assist readers whose mother tongue is not English or who are not readily familiar with the inner workings of concert details with a facile reading, explanatory or qualifying remarks are placed in '*parentheses*' next to the word itself. Meanings of specialized words are also given in the glossary section. Besides, there are analogies, contextual explanations, and anecdotal cases stated to elucidate certain crucial concepts. The picture collages help the imagination. Collectively, they should make the book worth an interesting read, and keep it as a source of useful information.

Besides providing entertainment and a 'novel like' drama, suspense, and mystery, the book strives to unravel certain hidden wonders of our environment & nature; Biological sciences & applied technologies; traditions & customs; joys & struggles; attitudes and lifestyles of humanity we witness in the world around us.

The book should benefit readers in very many ways!

Table of Contents

The first concert is being given by an enchanting young star Elise who recently burst into fame and recognition with her hit song:

"*Just say, I heard you!*"

The song is actually a Parody of "*Concerto d' Aranjuez*",(Concert of 'Aranjuez) the internationally popular '*guitar concerto*', by the Spanish composer Joaquín Rodrigo.

The palace & gardens of Aranjuez was built on the picturesque landscape by the delta of the 2 rivers in Madrid, Spain. Many ships might have set sails to different parts of the world for conquests and colonization from the very same docks of these rivers. Aranjuez is a site of a sprawling park planned and beautified as a resort for the royalties and the '*nobles*' with all the comfort and luxury. There stand a palace, pair of gardens, quarters with domed mansions, courtyards; convents; splendid promenades, and squares. Many stone carvings and facades depicting characters and figures from the Greek Mythology, and other ancient civilizations are installed there. The garden has been a major tourist attraction ever since. Invariably, one thing leads to another. To build, maintain, and enjoy all the luxuries, pomp, and lavishness, many things are wanted. It needed enormous wealth, slave labor, and other material resources. Colonization of unspoiled continents seemed to be the *straight* answer! Indeed, they chose the obvious option. For its exquisite architecture and beautiful gardens, it has been honored as a '*World Heritage Site*' ignoring its legacy and purpose.

The concert hall built here in Kochi has many parallels of grandeur and marvel to the aforementioned gardens of Aranjuez, sans the evils and atrocities associated with it. Elise's song is based on the singer's childhood event, which seems to have a significant impact on her life, development as a singer, and her anthropocentric (viewing and interpreting everything in terms of human experience and value) outlook on humanity.

People love to hear her extraordinary voice with octave ranges (F3-A6) from contralto (female alto voice) to 'coloratura' soprano (The ornamentation of music written for the voice with high sustaining notes, florid passages & leaps, including trills, riffs, and runs). This voice range is comparable to the Berliner "*World entertainer*", the great Ivan Rebroff who was credited with having a vocal range of (F1-A5).

The following is the theme and the '*muse*' (the inspiration that motivates a poet, artist, or thinker) of Elise's song. Initially, she simply scribbled it as a poem before it blossomed into a hit song. This is how it is read from her handwritten manuscript:

It was a long, long time ago!

In my childhood when

I was just a little girl,

Feeble, fragile, and timid

Were there in our own boats

Amongst many such groups

Moored right across the banks

Lined with tall coconut palms

Along the meandering canals

Separating green rice fields

Watching the fireworks

On the fiesta of St. George

At the Edathua church

With our respective companies of

Siblings, aunts, uncles, cousins

Sitting on the canopy

Singing, clapping, marveling,

Shouting words of wonder and excitement

At the spectacular sparkles

Sprinkled across the setting skies

And its wavy reflections

In the wavy waters below

You were cute and handsome

Flamboyant and jolly

Projecting an air of adventure

And self-importance

Wore a yellow shirt with blue stripes

Knickers with suspenders

Curly black hair

Neatly combed-back,

Jovial and joking

As I can remember with

Vivid recollection of the moment

That still lives in my mind

Suddenly, there came

A batch of thunderous cannon bolts!

The boat swung to one side

In the sudden shock, and exultation

From the people on the deck

It tipped, Rocked,

Tossed things overboard

Threw me off too

Along with the other objects

Deep into the reflecting waters!

The girl from the highlands

Timid of water and strangers

Was out of breath, sinking

Gasping and throwing up hands

In great danger and hopeless despair!

Alas! my own company of people

Enjoying the merriment

Of the roll and rocking of the boat

Did hardly notice!

Like a lightning, you jumped

Right into the water

Dived deep and lifted me

Pushed me ashore, gently

With a few long sweeps

That gave me precious air

To the banks, and to safety
Straight into the outstretched arms
Of my alarmed parents!

You made a gymnastic back-flip
Pushed off against the boat
That certainly jerked the boat
Alerting people on the deck

And swarm back to your boat
In elegant backstrokes
Looking like a folks' hero who just rescued
A hapless 'princess in distress'!

That's my last sight of you

Suddenly you vanished from view

In the darkness

That was falling fast

But not from my mind!

Of Course, not!

The following year,

Boldly, I came early

Moored at the same spot,

I scoured, searched, and asked

You were not there!

I was heartbroken;

Nevertheless!

Hoped you would come later

Only wanted to say my thanks

In person, and face to face

Instead, wanted to return home

And cry my heart out!

Your image is slowly fading from my mind

Also, the gravity of a certain danger

That I faced on that day

Fearing the real possibility

I scribbled it in my diary

Years passed!...

Suddenly now,

It springs out!

Takes a life and form of it's own

As a popular song

And this is my song...

My iconic song!

I ask the audience every time

If any had seen you or

Heard of you.

Anyone recalls anything

Anything at all!

Have ever seen you since?

You might have done many good deeds

Warmed and charmed many-a-hearts since

Just the way you did for me

With your kindness and valor,

As if it's your sublime duty

Without seeking reward, or recognition

Not even a word of thanks

Self-consciously, but selflessly,

So unique and spontaneous

Usefully gifted to you

In good grace and abundance!

I learned swimming since

Another 'repertoire' of survival skills

It helps to improve lung power

Stamina, curves, and figures

Clears sound cavities

Strengthens vocal cords

Toned and streamlined throat muscles

Helps my voice, music, and nimble range

The posture as well

Both enormously and

Very purposefully, indeed!

I can now hit up to G7 note

Fluently and Comfortably!...

Adding complex attributes

Runs & leaps; trills, riffs and flourish

The fluid range & extended notes

Voice control'n subtle vibrato and tremolo

An extraordinary gift I acquired

'Cause of you!

I say a little prayer for you

Recalling the true facts

Every time before my performance

This song renews my remembrance

Faith in humanity

It brought me fame and recognition

A fair living, and happiness

I am a singer

Only 'cause of you!

And I keep on singing

Imagining singing to you

In a nostalgic thanksgiving melody

Sending an emotional, longing plea

Hoping you'll hear it someday

Somewhere, somehow!

If you hear it anywhere

Simply whisper in your mind

"I heard you!"

That will be a real joy

A true recognition

Beyond all awards,

Accolades, and adulations!

God bless you!

My invisible friend...

The long-lost friend

My lifelong friend, indeed!

In her school, she won the prestigious "*Kunjan Nambiar Literary award*" for this poem, named after the inimitable poet laureate of the royal court of the great maharajah of Travancore. It was published in the yearbook at her college. One of her professors was quite moved and impressed by her story as well as how well she expressed her genuine sentiments. He thought it was too good to remain buried in an yearbook. So, he rewrote it as parody lyrics using the poetic formula by applying proper '*alliteration*' (repetition of similar sounding words starting with consonants); '*assonance*' (repetition of similar sounding words starting with vowels), and '*consonance*' (harmonious sounding words) to suit an internationally well-known nostalgic tune. Serendipitously, Elise made the song famous with her true sentimental feeling as well as her seamlessly complaisant (accommodating) voice and singing talents.

Today, on the occasion of the grand opening ceremony of this splendid concert hall, Elise is the chosen one to have the unique privilege and honor to stage the very first concert performance here.

The members of her large orchestra are taking their respective seats. All the guests and the audience are seated. Everyone is expectantly waiting for the ceremonious appearance of the talented performer of the evening shortly. That's Elise! She is going to start her performance in great style with her iconic song: *"Just say, I heard you!"*

Elise's concert is scheduled to take place at the new Amphitheater in Kochi, in the state of Kerala, on the first day of the week-long 'Grand opening gala'.

The state of Kerala, a region so green, fertile with exotic flora, fauna, fountains, spice gardens, natural beauty, lush pastures, mountain ranges, steep valleys, and bubbling waterfalls on the East; an 850 km long contiguous stretch of shores along the historically reputed Malabar coast on the West, thriving small towns and villages, tucked together in dense bunches in between in the mid-lands, is admiringly referred to as "*God's own Country*". Both the highest elevation in India outside the Himalayas- the *Peak of Anamudi* at 2700 m, and the lowest point *Kuttanad* at 2.25 m *below MSL* lie in the so-called Malabar coastal state of Kerala. It is a '*blessed*' tropical land of abundant biodiversity; 44 rivers, 2 vast lakes, thousands of long, short, large, and narrow interconnected canals for transportation, irrigation, and cultivation of a variety of crops, produce, and products. In the lowest-lying Kuttanad, near to the lakes and sea, there are many groups of vast and contiguous rice fields separated only by canals, unconnected by any bridge. They are mostly man- made, salvaged lands from the vast lake basin. They are protected from breaches of tide, water, and floods by wide embankments of mud extracted from the lake-bed itself only using manpower. The mud dike is anywhere from 1.2–3.00 m wide with flat crest settlement of mud and fortified only by '*downstream berm*' with minimal slope. A combination of perennial shrubs and ground cover provide some protection to the embankment from erosion, although they are not adequate against severe thunderstorms, and floods. The fields lie on the average 2.25 m below '*MSL*'. The peripheries are planted with coconut trees. They grow tall and strong within a few years and produce coconuts and many other useful products from the same trees during its lifetime of more than 50 years. It is the most resourceful tree in the world; appropriately and affectionately called "*Tree of Life*". In the Midlands of Kerala, there are rubber plantations. The rubber milk is a natural industrial product with many everyday uses and applications. On the very-high altitude plateau, the soil is composed of a mixture of sand, clay, silt, and organic matter. It is very suitable for growing spices. The farmers grow spice gardens, cultivating exotic, fragrant, medicinal, and aromatic spices that attracted traders and invaders to the exotic land from far and wide, and abroad. On the highest hills and mounts of the '*Hi-ranges*' there are several sprawling green tea plantations.

Despite all the endowment of natural resources, and above all, with a hundred percent literacy rate and higher education offering great human resources, the state is only moderately prosperous. The citizens value education as the utmost asset. There, only a handful of medium scale industries- a ship building dock, and a few small chemical companies operate. As the land is green and fertile, everyone hesitates damaging the environment and the pristine ecosystem flourishing here. Jobs are scarce. People are compelled to go to other states or abroad to find suitable and lucrative employment, where remarkably, they shine in their careers. Largely and literally, the productive land and the foreign money support the economy of the state! There are many colleges, universities, hospitals, agricultural farms, and businesses in the state. But they are seldom allowed to function efficiently. People who are left behind and less enterprising enter into politics. Politics become the predominant subject and persuasion that is for long now dominated by antisocial and unruly people. Its main objective seems to be meddling, disrupting, and wrecking everything using the motto of '*maximum power grab*'. The politicians are by nature and breed insatiable. They appear to derive sadistic pleasure in causing troubles, harm, and hurt to the people. They utilize the amassed power to extort money; steal and divert state's funds, and its valuable resources to themselves.

Beyond the annually recurring natural devastations caused by long and heavy monsoon flooding, and land slides, there are as much number of man-made catastrophes called '*Harthals*' (total shutdowns) that hinder everything and disrupt normal lives with total disregard for law & order, civility, or the grave economic consequences and human tragedies. They are organized by the ambitious and '*self-serving*' politicians with the affiliated gangs of hooligans; financed and supported by their respective parties. Out of utter frustrations, caused by numerous corruption scandals, government ineptitude, ignorance, pathetic attitude, and the lamentable conditions prevailing, the public quietly call it "*God's own country of the miserable Fallen angels*".

Quietly, they leave in search of suitable job opportunities. The majority of the people travel to the gulf countries looking for manpower related, semi-skilled, and skilled jobs. This has been a favorite destination of Asian expatriate workers for many decades. A few prove themselves worthy enough, rather fortunate enough for managerial and supervisory job positions. In this situation, both the expatriate employees and the host countries benefit. They settle for a long and symbiotic relationship. In some cases, both parties cultivate strong, cordial relationship, and goodwill. This story in this book, is one such case of beneficence, and philanthropy created in this fashion.

Then, there is the picturesque archipelago and the magnificent port city of Kochi at the '*embouchure*' (mouth of a body of water; also used to describe the mouthpiece of a wind instrument) of the Vembanadu lake, looking out into the Arabian Sea welcoming travelers from the land, islands, lake, sea, and air. The white sands on the shore, the lush vegetation on the land, and the tall coconut groves on the outskirts render a unique natural beauty to the sprawling lagoon. It may also be seen as situated at the estuary where the channel of the lake becomes wider as it approaches the sea, forming a reservoir, or a large '*mixing bowl*' of fresh water from the inland with the salt (tidal) water in the sea.

The charming and serene city of Kochi on the legendary Malabar coast of Kerala

They lend beauty, function, and purpose in numerous natural ways. There are passenger boats shuttling between the nearby island-villages. They ferry people, produce, cargo, and vehicles. The Arabs, Chinese, Persians, the Portuguese, Dutch & the British came. A few conquered the land and stayed there at different times and length of times of its fabled history for various reasons, but mainly to dominate international spice trade. Obviously, some harbored hidden, nefarious purposes, and intentions. When they left, or forced out, their ostentatious marks and monuments, as well as the scattered cultural remnants were left behind. As a result of the rich historical past, every state in India has become landscapes of forts, fortresses & towers; museums & mausoleum; Temples, churches & mosques; traditional and foreign architecture with carvings, mural paintings, geometrical & astrological shapes, and idolatry figures. The natives still preserve and assimilate them into the historical and cultural heritage of the land reminding the citizens of their checkered past.

Preserved colonial monuments

The invaders benefited handsomely from the plundered wealth, ancient wisdom, and knowledge drawn from the Vedas. They used the wealth and slave labor to build castles & cathedrals; palaces and Parliament houses in their lands; utilized their institutions, eloquent lies to justify and legitimize atrocities and genocides; enacted oppressive laws for the natives to obey. The invaders became the lords and masters. The colonies served as their marketplaces and get-away places for fun and exotic vacations. They forcibly took over the land; established farms and plantations; had the natives cultivate them; took away the prime quality produce to the international markets for greater profits; sold the '*glean*' (what is left behind after the main harvest or gatherings), and rejects in local markets. The greedy and unscrupulous insist that they always win in every step, turn, and time! In times of war and conflicts between the colonists, the locals were recruited as soldiers and infantrymen to fight in ditches and swamps; moors and mountains for the glory of the imperial crown. Such was the legacy and wretchedness of the colonial rule in India and everywhere else.

In the city, there are many things, sites, scenes, sceneries, and sights to enjoy. The traders /conquerors built several "*look out*' venues to warn of the new adventurers; erected '*observation towers*' at vantage points to enable spotting any spontaneous uprisings, or insurrections in troubled times. At other times, they served as sighting locations to view the many scenes situated the farthest away eyes can see.

Now everything is converted for the public use and civic purposes. The Observation towers are tall cylindrical structures overlooking the banks of various creeks, lagoons, and river sides. A carriage which can carry 30 people at a time travels up and down in a crawling, spiraling motion around them. Like the metro trains stops, they stop at every 15^0 turn for a few minutes affording unobstructed, 360^0 perspective of the city from the spot. The panoramic view is further enhanced by viewing through the special distance-viewing mirrors, prisms, binoculars, and telephoto lens mounted upon them. The viewing devices have many degrees of freedom and flexibility by way of pan, tilt, and zoom. Visitors get a plain, elevation, aerial, and ground view of the city; boat, cruise ships, loading & unloading of ships; the shiny shallow lagoon; road & pedestrian traffic; the edifices, park and meadows; houses of worship in contrasting scenes and glory.

The scenic port city has many faces; wield great importance in various fields

They may scour the skies too; spot special alignment and formations of celestial bodies. The faithful may look for the coming of the savior. On auspicious days, the place is crowded. It serves many purposes. Interestingly, no one knows what the other person is looking for. Everyone seems to be delighted and gratified.

Usually, the visitors and the tourists observe the sites from the beautiful garden parks on each side of the banks; the Marine Drive; the Dutch-built historic Bolgaty palace; the strategically situated Wellington Island, and from many boats and cruise ships. Each offers its own uniquely picturesque and breathtaking perspective. One cannot possibly miss the patrolling reconnaissance / surveillance Air Force planes, the Indian Navy ships, Coast guard patrol & rescue boats, merchant ships seen on the boundless, gigantic arch on the horizon; the approaching waves of the sea; the roar, force and occasional fury vested in them when the naughty dark monsoon clouds and howling winds ominously escort them lashing onto the trembling shores. The thick, long lines of tall coconut trees herald which way and how strong the wind blows. They '*rock*' and swing like the *rapturous* rock musicians, while the lightning sparks peak through them in intense bolts of lights. In the day times, to the joy and heartthrobs of sailors, they tell the marines where the nearest land is. The weary sailors long to set foot on the solid, grassy, friendly, dryland.

There are a number of long, bridges, and causeways built over the water bodies, connecting isolated wetlands, disparate landmasses, farming villages as well as Naval & Air force bases. The military bases vigilantly patrol the skies and the seas for coastal security; guarding against the return of the colonialists and the conquerors; foes and enemies; guns and missiles; pirates and terrorists; also sharks and sea monsters, protecting the population and the Republic. The painful memories and horrible events occurred in the past are indelibly imprinted in History and perpetually (impossible to erase) etched in the consciousness of the people. The enduring lessons learned in the hard way, shall never be forgotten! Prudent vigilance is simply a necessity!

In the ancient times, prior to the Common Era, Southern Arabia was a commercial center of spices such as Frankincenses, Myrrh, as well as precious metal, gold. The Arab mastery of the seas can still be gleaned from the "*dhows*" shuttling between Aden, Dubai & Iranian costs across the Persian Gulf carrying grain and finished goods. As far back as the first millennium, the Arabs came to the Malabar cost for the sole purpose of spice trade opening a direct trade route from the Red Sea port or the port of Aden across to the Arabian Sea. Their habit and affinity to tea is still evident in the customary greetings of guests and visitors by offering tea and drinking it frequently during the day.

Strategically protecting their trade interests, the Arabs controlled and extended their flourishing spice trade to the Mediterranean countries in the opposite direction as well. As a result, the trade shifted to the port of Alexandria which was already a busy commercial center. When the Romans became the rulers of Alexandria, they coveted their own domination in the spice trade and undertook regular voyages for the Indian spices. With the spread of the Roman Empire, the aroma and flavor of spices also spread with it. Seemingly, the scent of exotic spices was the sign of luxury and opulence, while the flavor the unmistakable symbol of ultimate taste. The Portuguese soon smelled and tasted the *Malabar* spices. They couldn't resist the temptation as they were bewitched by the prospects of a business opportunity and thus expanding their fledging influence. In those days, there was no bigger international trade than the spice trade, dearer than the present day oil-trade. Up until this time, the Portuguese were trading mainly in roots and fruits; seeds, and nuts. They came directly to the Malabar coast in 1497 with much more ambitions than mere trade. They attacked the traditional Arab traders at sea who monopolized the market until then. The Portuguese prevailed and became the first to establish trade route between India and Europe. The Portuguese were wicked, fanatics, and cruel. They interfered with the functions of the local kingdoms; instigated conflicts and wars between them; cunningly formed strategic alliances. Soon, Unilaterally, they waged wars against the maharajahs one after another. In the ensuing battles, maharajahs were defeated and merely became vassals of the conquerors. Having won consecutive wars, they turned on to the stronger kingdoms; achieved similar success. In the hubris and might of victories, they tried to forcefully convert people to foreign religions. The Portuguese were the first to come and the last to be kicked out of India.

In the enduring course of time, the Dutch, after forming East India trading Company came to the Malabar coast and prevailed over the Portuguese. The Portuguese sold their trade houses and other possessions to the Dutch and other entities; strategically retreated to Goa where there was little competition from mightier European kingdoms. For a short while, the Danish and French imitated their counterparts with similar tactics and '*modus operandi*'. But, achieved not nearly any significant or as much success.

The astute and adventurous British traders formed their own East India company and made their voyages for the renowned and fabled spice trade. Through a series of guileful and diligent manipulations, they acquired '*most favored trading status*' and other privileges from the Mughal kings in North India and regional maharajahs in the South. They grew the trade beyond spices to include sandalwood, bananas, sugarcane, and other '*cultigens*' (plant species that can grow only by cultivation; not wild), silk, textiles, and tea (they are credited with bringing tea and growing tea plantations in India). With the enforced exclusion from trading in Indonesia by the Dutch, India has gradually become the main trading center for the British. They expanded their related activities to other parts of India and gained influence; became enormously powerful; strengthened their maritime and military power; defeated many maharajahs and became rulers of the conquered provinces. They controlled, owned the entire national wealth, and resources, including gold, coveted diamonds, rare jewels, ancient arts, and artifacts. They were smuggled and transported to the homeland on their own merchant ships bereft of any account, provenance (authenticated document showing history of ownership), or assessing values on any of them. All the dealings were carried out in utmost secrecy, without feeling any shame, guilt, or dishonor. Showing no hesitation, but using the power of guns and canons, the British forced out all the remaining rulers and ruled India as a province of the British Empire in the evolution of East India Company to the ownership of the British crown. The guns gave them lands, wealth, power, domination, and control, so much so, it became the symbol of their might and glory. The guns are cherished and revered in the deep-rooted minds of the Western culture and lives. True to the saying, they value 'guns over the lives of innocent children'. The British presence lasted more than 3 centuries and the authoritarian British direct rule, the so called '*British Raj*' lasted just about a century. They were so oppressive and monopolistic that they didn't allow India to make, build, or manufacture anything; not even salt from seawater from the Arabian Sea, or spin their own locally grown cotton, and many other most mundane activities. The '*Industrial Revolution*' has started in England. They needed quantities of raw materials and markets for finished goods. The colonies served both the necessities! They confiscated sugarcane farms and sugar making industries; used it for making candies in their country and exported to others for profit. Their cunningness and skills are rewarded even this present time as a center of International trade for precious metals Like LME (London Metal Exchange),BIFA, (British International Freight Association) shipping, banking, charter business, so on, and so forth. They treated India only as a '*buyer /consumer state*' of British goods. They ruled with utmost arrogance, excessive penalization, and taxation; looted all the occupied colonies around the world and subjected them to their dystopian oppression; became very wealthy and lavish for the horrendous acts of world-wide thievery and plunder. The colonized countries are still reeling from the opposite extreme consequences! The '*Industrial revolution*' and the spread of Christianity were greatly spurred and hastened by the colonial conquests, wealth transfer, and exploitation by the West-European monarchies and countries.

Veni, vidi, vici" (the Latin phrase uttered by Julius Caesar, "I came; I saw; I conquered!"), the casually arrogant manner the notorious phrase was uttered, still haunt and resonate with the people. It is taken to be the callous mindset and attitude of the European colonial powers towards the rest of the world.

They came at the behest of their Kings and queens under the joint sponsorship and blessings of their hallowed religion repudiating the 6th commandment *(Thou shall not kill!)* of the solemn "*precepts & tenets*" of the "*Ten commandments*". Systematically, the colonists annexed the host nations to their own, without any intention of returning the '*conquered*' land to the native tribes! They, still occupy and own huge land masses of countries and continents of South America, North America, Australasia, New Zealand, and many islands giving the natives no rights to their own ancestral lands, nor freedom to speak their native mother tongue!

The insensitive and unscrupulous colonists came on the pretext of exploration, trade in aromatic & exotic spices, promoting goodwill and wholesome interaction between civilizations; bringing knowledge and technology; offering new moral philosophy, and good wishes to all humankind. Instead, after enjoying the friendly welcome and hospitality, they committed heinous brutalities of inhuman magnitude in appalling ingratitude; divided people; waged wars, ruled the land; subjugated the native population and infected locals with incurable diseases. The European colonialists plundered the wealth wherever they went. After stealing the treasures of knowledge and gold, they destroyed ancient civilizations and religions; impoverished the host nations; imposed their own will; spread their faith & religion branding the natives as "*savages*". They bought slaves from the colonized nations, forcibly used them to build castles and cathedrals, monuments and skyscrapers; roads, railway tunnels, and do hazardous mining & tunneling work, as well as work in cotton fields and tobacco farms under brutal climatic and working conditions, often failing to pay a minimum '*living wage*'. The conditions were exponentially worse than that of the Asian migrant workers in Qatar, building FIFA World Cup 2022 stadiums the media of the participating countries often lament and criticize!

From India, they stole coveted diamonds, the invaluable '*Kohinoor*', the blue colored '*Hope*', the priceless pale yellow '*Sancy*' and other valuables. The stolen diamonds, which were national treasures are periodically exhibited in European museums. They claim that

these were gracious gifts from the Maharajahs and Nawabs for their '*exceptional contributions and services*'! It were more plausible if they were to say "the ransom for sparing the dignity and lives of the maharajah's family". From African colonies they stole prehistoric relics, fossils of the earliest geologic age.
Why in the world certain people only receive expensive gifts, but never have to give any! They only sell goods at huge profits to the hosts.

Apparently, the multifaceted, colorful jewels were not a good match for the skin color and for that matter, didn't look good on the queens and princesses of the Maharajahs!
Besides, the precious metals, they stole unique articles and artifacts that are seen and made no more, such as copper tools, ornaments, art works, embroidery, even tusks of exotic animals, and of species long extinct centuries ago. One may imagine the value of these collections comparing each item to the recent auction of the most expensive painting "**SCREAM**" by the Norwegian painter Edvard Munch.

Astonishingly, the conquerors and offenders who came with guns and ammunitions were finally ousted by prolonged non-violent struggle of "*noncooperation & civil disobedience*" lead by Mahathma Gandhi and through peaceful international campaigns. It is one of the three such successful freedom struggles known in modern History! The campaign against the '*apartheid*' system of racial segregation by the South African people, lead by Nobel Prize for peace winner Nelson Mandela also achieved the same feat (an act or accomplishment of great courage & skill). The Civil Rights Movement in the USA has also been a non-violent struggle. In contrast, in the case of South America, it has been a total surrender and capitualation; an absolute and perpetual subjugation, and colonization.

In India, now, the nation building has been in full swing. The tasks associated with catching up with the long-lost centuries of colonial tyranny and oppression, as well as the consequential inability for '*self-determination,*' are indeed daunting, and arduous. The sick feeling of collective '*inferiority complex*' has been replaced with a '*proud sense of optimism*'. All major issues are set aside, as freedom, national advancement, and self-determination take a firm hold among the population. It prevails over everything, despite the extreme radical policies of the current '*ultra- radical religious*' ruling party.

Infrastructure developments in India

In the suburban proximity of the city of Kochi, lies the serene *Cherai* beach in *Vypin* island. It offers a breathtaking and inimitable view of the sea and the sunset in its actual pomp and glory; its shallow waters beckon for an adventurous swim; the soothing cool breeze blows away perspiration; soothes the torrid humidity and heat. The inexorable sunlight, and the ever present salty breeze, offer a sun-bath for tanning of the skin as a free gift from the nature. A well manicured '*esplanade*' (a flat open stretch of pavement designed as a promenade along a shore) with flower garden and exotic shrubs provides a contrast, some shade, a place to sit and lookout from. A great opportunity shall not be missed to witness the true Arcadian (rural,rustic, pastoral settings suggesting simple lifestyle and contentment) lifestyle of the local people in their natural ways, with their nominal needs. They take little; consumes only the essentials. They are '*minimalists*' in the true sense of the word. A beach devoid of skimpy bikinis, is a social flocking point or meeting place for an enjoyable evening stroll for the family, friends, and others. The visitors watch women dressed in full length colorful Saree & matching blouses with flower tops as their hair decorations, replacing fancy hats; men in white loincloth and Juba (dhoti, lower & upper garments) let the sprawling shawl hanging from the shoulders flutter around as a kite trapping some cooler air that is directed towards the perspiring body, saunter the beaches. Some, curiously watch the many body languages of actions and reactions; the typical '*head oscillations*' of the people expressing agreement, disagreement, or acceptance, distinguished only by its *incidental* variations. The people are spontaneous; conduct, and act out their ways and wishes in the most natural human ways.

A natural and descent beach scenery; nothing shaming, nor too ostentatious to blush; a normal, and peaceful life

Yearly floods and storms fail to discourage or dislodge the inhabitants from their habitats. They are closer to their fellow beings and environments than to machineries, computers, or gadgets. The islands and coastal areas are the first to be affected by the perils and furies of nature caused by the man-made global warming. These people cause none of it, neither reap any economic benefits from its comparatively short-time gains. The island dwellers all the world over, are the patriotic guardians of the environment. They are rugged & resilient; perseverant & hardworking; possess an indomitable spirit of survival. Their endurance and life experiences vouch for the fact.

The food vendors prepare steaming hot items in the open air with aromatic spices. It's so appetizing and makes people want to have some. Visitors begin to flock to the food vendors though it's some further way to go before being fit for eating. It reminds me of the story I heard. During World War II, the defending Russian soldiers set up field kitchens where the advancing German troops were bogged down and immobilized, having experienced severe logistical problems of food and fuel in a winter blizzard. The ice and extreme Siberian cold is so brutal; "***hypo**thermia*" (below -30 °C) is fatal! [In contrast, ***hyper**thermia* occurs when body temperature rises above 39.5^0 C]. The hunger became even more unbearable by the aroma of cooking from the enemies. Jealousy, hunger, and the extreme weather combined, and conspired. It caused a telling effect on the outcome of war itself!

Politely, the vendor/ owner / manager informs the patrons that food will be ready soon without giving any time frame. Dejectedly, some leave, and go back to what they were doing before. The cooks sprinkle home-made mix of curry powders in different combinations and colors after each act of tasting. It seems that the optimum taste constantly evades them! Leave it at that! Salt need not be added; the air supplies genuine *sea salt*! It's quite different from table salt that contains additives and preservatives. Sea salt carries rare minerals and nutrients. In moderate level of consumption, it can provide several health benefits: soothing sore muscle, improve teeth, skin & hair health, blood sugar regulation, balancing fluid levels, correcting distribution of the electrolytes of sodium & potassium (they form *oxides* that dissolve in water to form alkali) cardiovascular health, etc. It is the salt, a venerable symbol of India's freedom struggle against the colonial Britain that was taxed in order to prohibit making in India in favor of the British table salt in bottles! The "*Salt Sathygraha*" marked the beginning of the "*Civil Disobedience Movement*" that gave impetus to the struggle in toppling the brutal British rule. Indians, therefore owe much to the sea salt!

The mouth salivates, and craves for a bite. Tantalizingly colorful dishes of sea and lake-water fish, fresh from the day's catch, topped with richly sautéed onions in *masala mix;* tomatoes, hot chillies, garlic, and vegetable slices cooked in coconut oil, are enticingly arranged and displayed to the full view at different levels on the stalls. If you taste any, 'um....um.....yummy!; you'll want to have more of the same, and the other kinds as well, just for a bite and taste! The vendors generously offer samples on toothpicks for simple tasting. The hungry are befuddled, as all the items taste tantalizingly delicious. The breeze spreads the pungent aroma of the fully cooked spicy food items; aerially propagates the message without words, "folks, the food is now hot and ready; come and get it!" reminding of mothers' call at dinner times. The hungry re-assemble at the stalls; survey what's on offer and compare prices. They cannot make up their mind; everything is enticing; decide to go by the look and price of it; randomly select; order more than they really need. The recommended amount of calorie intake an average human body needs depends upon age, gender, body size, and living environment. More scientifically, it is based on *a.)* basal metabolic rate; *b.)* physical activity; *c.)* thermic effect of food. Or simply and roughly in numbers: less than 2000 calories /day. [***basal metabolic rate*** is the amount of energy needed for the body at rest to carry out the basic functions of living which is the most amount of energy-consuming component; ***physical activity*** is the second-highest component and vastly varies, but largely related to the body weight; ***thermic effect*** is the calorific energy needed to digest and breakdown the consumed food into basic elements to enable absorption into the body]

The flocks of crows scavenge on the spilled food particles; keep the place tidy and clean. The ubiquitous, furry squirrels gnaw on the nutshells, scurries away with the excess, make burrows in the sand; hide them in the holes, or on trees for future use. As we can see, the '*Sciuridae*' is prudent and provident mindset! However, later, they cannot seem to locate the same spot that were soon made obscure by the human traffic, and under varying sizes & weight of footsteps. Look at the poor animal plights!

The fellow species and predators watch them and steal from their hoard. In this world, one who sows don't always reap! Stealing and many other vices, seem to be the universal '*animal/human natural instincts*'. The list of others is mentioned in the "*Ten commandments*" purportedly revealed to Moses directly from the heavens. It is high time that the original plaque written by "*God's own hand*", and kept in the "**Ark of the covenant**" be brought back from its highly secretive location somewhere in the Ethiopian mountains, and revealed to the public once every now and then to inculcate the folks of '*good conduct & living*'!

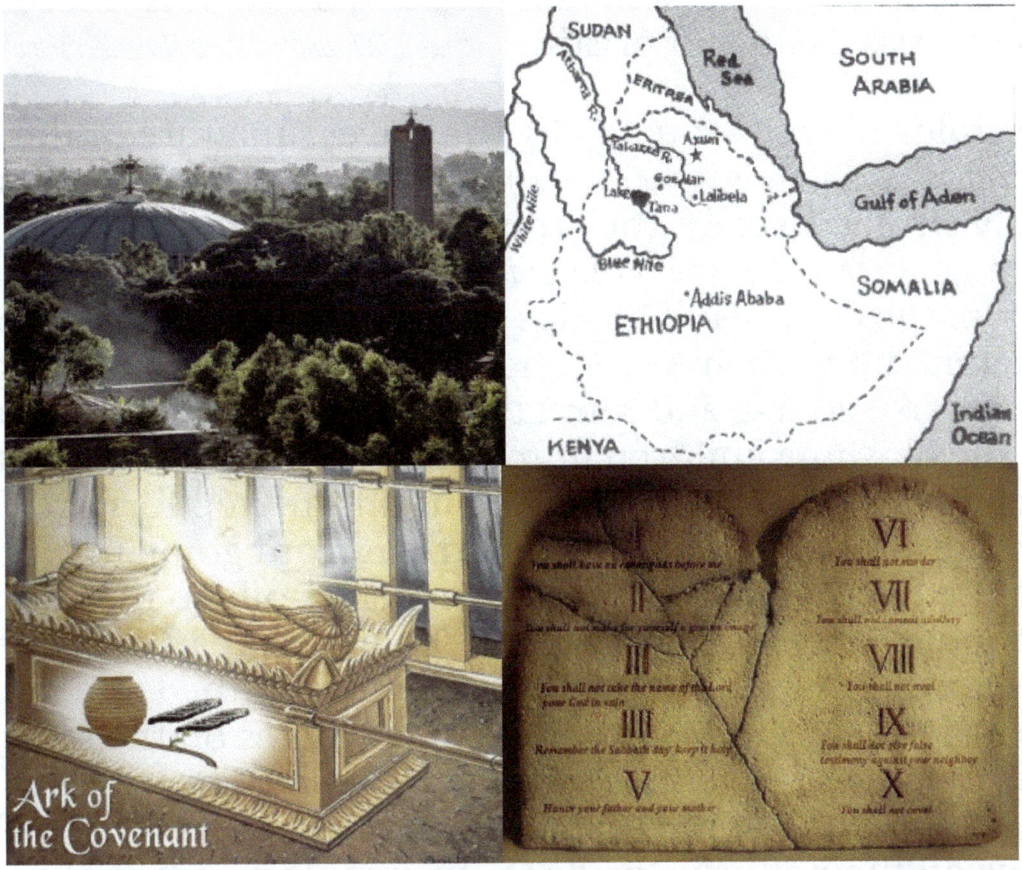

The tablet of Ten Commandments, an important content of 'Code of Conducts' Contained in the 'Ark of the Covenant'!

On the beach, when the waves approach the shores, its crest of cascading waters and foams shines in the colorful glow of the setting sun. As they strike the shore, expose the eggs the turtles had laid on the shores in the '*third quarter of the waning moon*', the period of '*low tide*'. They had been snuggly buried in the burrows. Many, now, lay scattered on the shore inviting birds to devour and feast on them. The bounty of the sea is largely hidden deep and far away from our view. They contain rich nutrients and exotic food for all species including ourselves. Perhaps the life itself started from the sea, from the 'Tardigens /Tardigrades' (a.k.a water bears. Defying its bombastic name, they are virtually indestructible micro creatures with astonishingly higher survival and endurance qualities to all other species!). The sea act in many other important and mysterious ways; especially determining the climate and weather. No wonder, the meteorologists call the *weather* which is referenced to '*over a short period*' as the '*mood* ' while *climate* which take into account the conditions for a longer span of time, the '*character*'! The essential lesson is that we must preserve all natural resources clean and pristine; do not pollute and contaminate indiscriminately!

The children are fascinated by the waves, unaware of the dangers of '*rip currents'* (a strong flow of water running away from a beach, back to the open ocean) that are deceptively strong. Usually, they travel very fast and away from the shore cutting across the '*breaking waves'* (breaking of surface water on coastline) in its hurry to join the deep sea. The kids dare and chase them when they come reaching up to their knees and within their grabs. The waves suddenly reverse course and retreat defying the catch, toying and mocking the blushing children. It makes them only bolder and defiant to dare the waves by moving a brave, but unsafe step forward. The parents chase after their kids; restrain them from going too far into the deep. But the few angry ones shake themselves off; go their ways and make yet another brave step forward. The parents can only keep a vigilant watch and stay close by. They are fully clothed and poor swimmers themselves. Some children scrawl lines on the sand and challenge the waves to '*cross the line of forbearance*'. The waves menacingly approach to the lines and cowardly go back. The children gloat; show off their effusive pride to the people standing around. They taunt the waves by challenging them; move the lines forward. When the boundary lines are finally breached, they look embarrassed, and hurt. In another spot, no sooner the children finish building their towering sand castles, than the naughty waves approach; cruelly wash them away. The children chase after the waves in anger; throw sand at them. Some others feel helpless; cry loudly, staring at their empty '*autonomous plot*' of land which was their pride and precious holding only a moment ago. It's the first entry in the '*damage & loss*' column in the account book, that they do not know at this time that they have to update frequently. The toy vendors and other peddlers pass by; hawk their goods; elicit inquiring looks; hotly haggle; finally sell items at a huge discount for half the original asking price. Parents buy them to quiet down their crying children. They are also quite satisfied with the '*deal*' and the peace it did make. The children forget their misfortune and turn the attention to comparing their toys with that of others, before starting to play with their own.

The land strip offers differing views from the same spot. A striking dichotomy (A division into two contrasting parts or categories) of rural and urban communities become apparent at short distances. When looking towards the city across the lake, the hi-rise city buildings & landmarks can be seen in the fuzzy distance as a background; lush greenery and reflecting waves in between. Dominating edifices and religious icons of places of worship dwarf other structures- cranes and billowing chimneys. The spires of churches, the '*gopurams*' of the temples, and the minaret & crescent of the mosques peep out of the skyline, heralding the communal demography and religious zeal of its inhabitants. Towards the sea in the opposite direction, it is a starkly contrasting view of '*nothingness*' or the world too far beyond the range of human vision and clear senses. Sunset is the only spectacular event taking place at the distance, except the glimpses of fishermen's boats arriving the home shore in the very early hours with their bounties of fresh '*catch*'. Obviously, they are tired after the nightly adventure; nevertheless, very eager to reach home in a hurry and catch upon the deprived sleep, soon after a cool bath that washes away all the fish-smell and sea salt off the body. This can happen only after a satisfying breakfast lovingly served by the most caring and attentive wife. The coming home sight is highly exciting and expectant for the fishermen-families, as well as the many vendors who have gathered and waiting even much earlier. The many scouring eyes try to spot the first glimpse of the oncoming armada, as soon as they emerge over the horizon. It takes a while for them to reach the home shore. But regardless, a flurry of hustle and bustles occur almost immediately. They have heard about many tales of rough seas, nightly thunderstorms and the sea monsters. Collectively, they heave a sigh of relief; give out a '*shout of exultant joy*' of surviving yet another night of an adventurous fishing expedition. Up on the arrival of the boats, the vendors load the fresh catch up on the many lorries that are lined up and depart as soon, to reach their markets far and wide, before its openings. The cycle loaders get the next chance followed by the foot peddlers. They settle the prices later. The seller-buyer relationship is a long enduring business pact and based on mutual familiarity, trust, and respect. They have a long business history together, some going back to a few generations.

Pretty soon, majestically, the sun ascends the sky; stands straight up vertically above the '*meridian*' and over the equator (noon, reaching the highest point). It's become unbearably hot. The waves pick up additional thermal energy and kinetic energy. The conditions steadily grow severe to unbearable. At this time, the beach is completely deserted. The tireless wind blows escorting high energy waves to the shore. The dense waves pack several forms of energy. They lash and sweep the coast; deposit many microorganisms, debris on the shore before retreating in a haste. Often, the sea becomes angry, agitated, and poses several threats. They include extreme natural events such as hurricanes, coastal storms, tsunamis, as well as longer-term risks of coastal erosion and rise in sea level. In the afternoon, the sun progressively descends; causes slight abatement of heat. After having lunch, people at home feel a bit dozy and drained off energy; often slip into a siesta, or nap. As the sun descends further, it is seen growing in size and coloring the skies. It casts '*halo*' around the nearby clouds giving them differencing transparencies, and visual properties. The much anticipated evening has arrived! It's the best time of the day for everyone, though the period is rather short. The schools are dispersed. There is hustle again in the neighborhoods. People come out and survey the surroundings and conditions. When both the heat and humidity become tolerable, they stroll up and down the beach; find friends, and neighbors; chat, and laugh; find the late evening atmosphere quite pleasant. The vendors arrive much earlier and open their businesses. Children play games in the sand. The beach is their football, volleyball, and cricket grounds. They bowl without pitching the ball. A batsman smashes a '*full toss*' with a swashbuckling swing, attempting a Sixer. He gloats, but a fielder comes from nowhere and snatch it from the air in '*hop-step and jump*' style and in a final acrobatic dive. The fluffy sand dunes provide protection for the crash landing. The batsman makes a dejected departure. A section of the spectators run into the field; join the euphoria with the bowler and the team of fielders. The umpire orders them away. The intruders argue, but finally heed to the *superior power* and retreat. The others mock the batsman who walks back head down and eyes planted on the dust of sands in front of him. The game resumes as soon as the next batsman takes the stand. The same time, a goal has been scored in the neighboring football field. The spectators rush to the spot. They rue missing the spectacular goal; hear an account of it from onlookers; inquire if anybody has recorded it on their smartphone; remain there hoping another goal would come soon.

During this time, the sun falls rather steeply. The horizon becomes delightfully colorful. The water below reflects light making it an even spectacular spectacle. Moments earlier, another set of visitors and admirers have gathered at the beach and patiently waiting to witness and enjoy that spectacular daily event of sunset when the sun kisses the Arabian Sea an intimate and solemn "*Good night, goodbye*" farewell. They marvel at the spectacular color patterns displayed on the horizon. Some clouds drift away unmasking even more patterns and underlying clouds of different color combinations right beneath. The joy and excitement of the onlookers are clearly palpable. They wait until the sun has fully immersed in the sea and disappeared from view casting only a faint hallow around the colorful clouds immediately above. The nosy children head home chattering. They are required to be home before the time of nightly prayers. Immediately thereafter, they must do the homework. Their sisters have been helping mother with the chores and probably finished their homework too.

At this time on thc bcach, everyone feels a sense of peace and quiet. The night has fallen! Afterwards, people enjoy a peaceful stroll along the beach in the salubrious breeze listening to the gentle waves stroking the shore. The battered shore needs an occasional massage. Little children with their parents become excited to watch the flocks of birds departing in some haste along the shoreline towards their perches far away. They form delightful formations on their way back home to their roosts. Soon, everyone on the beach will also make their homeward journey leaving behind only the lonely hearts and secret lovers. The lonely may talk to the waves; shed some tears; the longing hearts dream and imagine under the milky moonlight that dim the twinkling stars up above the sky. They wonder what may become to their destiny and desire!

The growing city that shines in many electric lights is visible through the thick darkness at some distance showing its alluring charm, beauty, and sophistication; intelligence, passion, and desire to ascend to great heights and achieve even greater glory. It is the home of education. Many specialized colleges, universities, and Research institutions; Centers of Science, Mathematics & Engineering, Humanities and Arts; also a conglomeration of Info parks specializing in Computer & Digital Sciences, Internet & Telecommunications, and technologies in many other fields are situated and functioning. It is a cosmopolitan city of spirituality, business, banking, shipping, and trade. The all aspects of life, work, and human advancement mingle in '*contrapuntal*' (A melody added to an existing one to provide harmony whilst each retains its simultaneous identity) harmony, necessary cohabitation, and tolerance allowing all faiths, religions, and creeds both ancient and modern to flourish alongside one another. The city hosts many places of worship: temples, mosques, churches, convent & seminaries, ashrams, and madarasahs claiming its status as a spiritual center as well. There is also a synagogue, the '*Paradesi Synagogue*', the oldest active synagogue in the entire *Commonwealth of Nations* which also peacefully coexists in immediate proximity with a Hindu temple, and a Catholic Church in the midst of other mainstream entities.

The Jewish community in Kerala which is now reduced to just one family belonged to the '*sephardic*' tribe who had inhabited and thrived in the Iberian Peninsula (Spain and Portugal). The Moors from the name '*Mauritania*', a group of Muslims and Berbers from North Africa, presently the countries of Algeria and Morocco have been occupying and controlling the most part of the Iberian Peninsula for about 700 years. But following the marriage of Isabel and Fernando, the two Christian Kingdoms were united and jointly fought against the occupiers. The Moors were defeated in the Middle Ages. Subsequent to the victory, the religious fervor was at feverish pitch, and the brutal period of the '*Inquisition in Spain*' (15th century) started. Tribes who were not identified as Christians faced persecution for certain, culminating in their expulsion. Like many '*unrecognized*' tribes, the Jewish tribes also fell victims to this ordinance. Thus, the Sephardic (those from the Iberian Peninsula) Jews became isolated, dispersed, and ordered to leave the Peninsula. The expelled took different destinations and met with different destinies typical of wandering ethnic groups.

Following the exodus, some set sail in flotillas; the direction of the wind determined their destination. It landed them at unknown islands and land masses of the '*Maghreb*' states of N. Africa. But no one admitted them; sent them away considering neither their plea, nor plight. A few kind folks and communities offered them food and provisions. They wished the drifting visitors "*bon voyage*" and "*goodbye*"; send them away to seek their own fate. On their journey, they encountered perils; battered by storms; drenched in torrential rains; fell sick; suffered hunger, and disease; endured several inhuman conditions; lost natural immunity, and strength; sadly and inevitably, many succumbed, and perished. Their boats sailed aimlessly in the vast seas. Ultimately, they were found by the deep-sea fishermen off the Malabar coast; escorted them to the coastal town of Mattanchery, Kochi. There were approximately 500 people in the flotilla. The great humanitarian Maharajah of Kochi, HH. Rama Varma moved by their wretched state and sufferings from the long, arduous journey, granted them asylum, assistance, shelter, permanent refuge, and a land near to his own temple to build their place of worship. The same spirit of '*religious accommodation and mutual respect*' the great Maharajah has created permeates and endures to this day among people of his erstwhile kingdom state! Thankfully, in the history of "*Bharatham*" (the greater India), except for the Mughal dynasty, all the rest of kings & maharajahs had been benevolent and caring of the citizens they ruled and the humanity as a whole. Magnanimity is a virtue in Kerala!

The secular religious icons stand in proximity in the spirit of mutual respect and coexistence, a common site in Keralam
A sight similar to the '*Town of Jerusalem*', but in stark contrast in all other respects

Well, this is not an isolated case in India's history of granting religious refuge and permanent shelter to the persecuted. Persia was a center of civilization, influential, and possessing of advanced knowledge. In the early 7th century when Persia was conquered by the

Arab Muslims, the Farsis belonging to the second-oldest monotheist religion "*Zoroastrianism*" were terrorized. Many fled to India and sought sanctuary. This flight and migration was partly land based; secondly by sea routes. Since the times, there are more Farsis in India than anywhere else in the world. They flourished and became prominent citizens, business people, and industrialists. I had never heard about this community until I went outside Kerala searching for a job. My first ever job, as an *apprentice engineer* was in Poona, Maharashtra in a Farsi-run Industrial electronics company by the Poona-Chinchwad expressway. I mistook it for a British company, as the company, as well as all its top executives had British sounding names. They spoke fluent English; dressed like them; wore stylish hats, though not behaving as much snobbish or nonchalant. They were fair skinned too, that consolidated my notion. I assumed the Shah of Iran was Farsi too. One of the three fabled kings who traveled to Jerusalem to visit the newborn Jesus was believed to be a Zoroastrian king!

Commonly, in the world today, there are caravans of people belonging to various ethnicities, religions, and minority factions flee their beloved homelands due to causes of existential threats of climate change effects, war, famine, economic hardships, discrimination, ethnic cleansing, violence, terror, and brutalities. It's a heart-wrenching sight! Humanitarian organizations and people rescue them. Many countries grant them temporary shelter, refuge, and assistance. The major reason is the frequent natural catastrophes caused by the global warming, and the consequent cataclysmic *climate changes* affecting every country on earth, but most severely the island nations, although they are the least and the last people on earth who could cause this ominous disaster. The agricultural farms become inundated in one season; suffer severe drought in the next; many-times the both extremes one following the other. Their crops are destroyed. The salty seawater render them barren, unsuitable for cultivation for a long stretch of time. Devoid of moisture, the dry land warps and cracks; fissures widen; woodlands burn scorching the earth and dwellings, forests and pastures; Savannah and fields. Ice caps and glaciers melt in the severe heat, raising ocean, seas; flooding rivers and canals. Storm surge, coastal erosion, hurricanes, and typhoons lash across land and sea, wrecking things, disrupting life, and killing masses of population. Their cows and cattle perish. The spared but the hapless, and starving population are forced to flee. For them, a known landmass is far away indeed, and reaching there crossing the treacherous seas in between is unmistakably a formidable undertaking. Many perish in the unseen breadth of the brutalities of the land and seas. If and when they finally arrive at an unknown foreign land, they face unwelcome stance. In a few countries, the walls may fall while in several others, new walls swiftly rise. It is by no means a kind or a compassionate world as the religions constantly preach, but a selfish and fiercely competitive world.

I was told the story of the Jewish exodus by Basil, my best friend when studying at Maharajah's college, Ernakulam. I've never heard any story so ancient, graphically narrated by a person to whom it was told first-hand by successive generations of own trusted forefathers. After many decades, I still recall vividly when he told me that his family has decided to leave for their ancestral land of Israel, which was newly formed and was in its infant state of becoming a sovereign nation. They felt obliged to leave and contribute to the event, also creating self-identity. With spice trade taken over by the very aggressive Bombay based Gujarathi '*trading houses*'; their community was dwindling and diminishing in size and prospects; also being increasingly isolated; found little chance of survival here, and are again forced to leave. However, this time, it is not by threat, persecution, or expulsion; on the other hand, on one's own free will, made under the duress of dire natural circumstances of totally different kind!

He appealed to me not to feel offended that he or his community were ungrateful for such a gift of long and generous hospitality they have accustomed to enjoying all along from the Maharajah, his subjects, and the citizens of the state. He added that his community will forever remain immensely grateful that the then Indian Prime Minister of India, Mrs. Indira Gandhi had come to inaugurate the ceremonies commemorating the 400th anniversary of the synagogue.

The scheduled departure came both as sudden and shocking news to me. It was delivered in sincere regret, with feelings of shame, guilt, and sorrow in equal measure. It was a sad farewell, as one can imagine! Basil was my classmate. Somewhat like himself, sometimes I also felt like an '*outlier*' (a person situated outside of the main body) experiencing a form of isolation from being brought up in another part of the state (Travancore) and with sightly different speaking accent, and mannerisms. In those times, it was enough to sort out people from the mainstream. Occasionally, after the team sports we played, we used to relax at the beautiful Subhash Bose park situated right across the busy street from the college campus at the edge of the bustling straits, looking towards the Wellington island, the home of the naval base, wharf, shipping, and business center.

The Jews were expelled many times, from different countries at different times in History. One such event was the subject theme of Giuseppe Verdi's famous opera "*Nabucco*". The '*plot*' was the plight of the Jews, after their exile from their holy land of Jerusalem, following the conquest by the ruthless Babylonian King Nebuchadnezzar nicknamed Nabucco. The "*Chorus of the Hebrew slaves*" is the famous chorus describing the abject grief, utter despair; among them a wishful, but deceptive longing and hope of returning to their homeland some time in the future. The beckoning memories of the city of Jerusalem, the destroyed '*Temple of Solomon*', Galilee, its surrounding hills and plateaus; the flocks of sheep they left behind; their friends & relatives they may never see again, haunt them during their captivity. The opera and song are memorable classics, strong in emotional and musical expressions. It is popularly known as Va pensiero which brings out intense and enigmatic feelings of the exiled; a melancholic song I have finally learned to play and share. I am reminded of Basil and his final words of farewell, fraught with sadness, ambivalence, and fear of uncertainty for his future! If he hadn't told me the story, I would not have known or recorded this song. I haven't heard anything about him since that day! In the melee and efforts to make a respectable career, it never occurred to me to make inquiries or attempts to contact Basil. But I did inquire and read about the ancient Synagogue. I was told that a single family is still surviving there. They are obligated to stay for the sole purpose of taking care of the place of worship and perform religious rituals. For their continued existence and welfare, the family depends heavily on a kind, trustworthy Muslim neighbor who assists them in managing many non-religious matters.

The noble legacies and wonderful settings of the city, flattered as the "*Queen of the Arabian Sea*" provide a suitable location for the new Amphitheater. It is situated in the middle of a sprawling green meadow on the banks of a long canal. The fairly flat meadows are aesthetically partitioned and lined alternatively with flowering, non-flowering shrubs, and fragrant shrubs in beautifully arranged blocks. Exotic flora, a pond with tropical "*floating pond plants*', and water fountains oozing out water jets in distinctly different arrangements and shapes surround each building. One location has a dynamic rotating nozzle. It dances; swing side to side; jumps up and down, and spin at different speeds in its routine play. When colored light beams fall on them, color drops carrying bright reflection are sprayed around in a stunning spectacle. At another place, a dragon spit thin sheet of water stream from the "*laminar fan jet*' fitted in its mouth. At a third place, a '*bouquet*' shower jets out water in the shape of flower petals falling fast in the wind. There is also a '*Geyser*' fountain jet shooting out several columns of waterfalls in a row as a cascade (a series of waterfalls). Another one is the '*umbrella /mushroom*' jet whose circular nozzle, facing upwards is held on a pedestal high above the water level. It sprays water continuously, smoothly, and evenly around in a circle giving the fountain the shape of a '*parabola*'. The last one is the '*Sun /Ball*' spray. Multiple nozzles are arranged in a spherical pattern dispensing water at many angles with great precision. At night when the lights are projected on it, multiple virtual visual illusions of '*solar activities*' can be observed. The settings help to present an aesthetic, exciting, biophilic, and natural environment in the Art Center complex.

The interconnected rubberized *walkways* are shaded with solar roof structures. The shape of each differs as flat, '*Inverted V*', slopped, rounded, to optimize solar power harvesting. Battery operated diffused lights are activated and controlled by a dedicated light sensor for each section. They illuminate the walkways.

All utility cables such as electric, fiber-optic, multi-fiber, communication, RF, and heavy-duty types are laid underground in mutually insulated and separate conduits. There are a few individual satellite dish installations for the "*purpose-delineated*' services for Television, data, cloud, audio, video, signaling & metering, and for control signals. There are also large size, trackable uplink satellite dish installations to upload signals to various geosynchronous (geostationary, so that both the satellite and earth are in same position relative to each other, appearing stationary) TV & communication satellites. There are redundancies and contingencies built into every design to assure reliable services and operation.

The main and the largest building on the complex, is the Art center concert hall itself. It is situated majestically, flanked by other auxiliary buildings of various sizes and architecture. For the required 3 D acoustic excellence, the Art Center is roofed with a suitably shaped dome and an enclosed circular structure, despite the fact that the term *amphitheater* almost always means an open air theater, without shade or cover except for the stage alone.

In all other respects, it meets the standards of designation. It's a gigantic circular structure 110 m in dia. comparable to the width of a FIFA approved football field; 1/3rd of the space is occupied by the stage & the backstage. The rest is available for seating purposes in a combination of stories, balconies, and mezzanines of various shapes, and seating capacities. Its large many faceted atrium and other entrance hallway extends 9 m further outwards with ornamented, sculptured, colonnade of Corinthian columns with glazed marble and elegant mezzanine floors. There are symbols, icons, gadgetries, and other elements exhibited here at this location. The space lends a chance to adjust our innate *biophilic* (A strong emotional attachment to the actual living environment) senses accustomed to nature and environment to an entirely different world of advanced technical features, splendor, refurbishments, and settings offered in the hall.

An escalator takes VIPs, celebrities, and special guests to the various levels and locations of balconies with additional comfort, security, vantage points, and amenities. The *capital* of each column is engraved with Vedic figures. The walls carry inscriptions of verses and words of ancient sayings, wisdom, and prophecies. They carry important reminders, concepts, precepts, and tenets to make a nation cultured, civilized, and for the purpose of moral implementation. Nowadays, lives and culture are threatened by guns, violence, drugs, crimes; depiction of nudity & vulgarities, and foul languages. The writings must serve as an '*antidote*' for these modern day perversions and depravities helping people to maintain sane, serene, stable, and sound values.

The entrance vestibule is a glittering place with electronic imagery, display panels, virtual studio arcades; concierge, & reception desks; an open space to look out at the natural world and relax. There is yet another area with well-appointed and stylish furniture'; children's play area and a row of the all important ticket counters. The walls display large enticing posters and trailers of the upcoming events. On the mezzanine floor are business offices dealing with concert hall matters, service, operations, and related matters.

All in all, it's a splendid place. The airport-style icons and sign designations lead each visitor to the right places on one's own. The *interactive boards* at 'T*urning points*' guide a confused visitor showing the '*right way*' by visual and aural instructions. Also, one can upload the same instructions on to the smartphone by '*touch mode*'. There is a honest attempt to make everything autonomous and efficient, yet hassle-free!

Chapter: 5 The Sheikh, the man behind the enterprise

Today, is the opening day of the new Philharmonic & Concert hall built as an Amphitheater with 20,000 seat capacity. It is the only kind of its size and purpose in the country. It was built by a consortium (an association of businesses, financial institutions, and investors to undertake a Joint venture) founded and funded by a philanthropist '*Arabian Sheikh*'. He was greatly impressed by the numerous contributions and services rendered in various fields of Healthcare, Education, Business, Finance, Accounting & Banking, Aviation, Agriculture, Hospitality, and infrastructure /oil industries by the people of this tiny state of Keralam to his native country. He felt obliged to reward them in a grand gesture and with utmost gratitude. He is also said to be planning to build a state-of-the art Football stadium in Kozhikode, another city in the same state.

This Concert hall has been an ambitious project, 9 years in the making at an estimated cost of $195 Million. The Sheikh assembled and constituted an impressive international team of renowned experts and designers in each of the specialized fields. It contracted Rolta India for the geodetic survey, 3D modeling, digitization/ Net Centricity /Synchronization, Coordination/ Execution of the project management, Design, Construction, and Operations/ Logistics; Pininfarina Mahindra group (originally a sports car designer) for an aesthetic and exotic building design concept; structural designers from Berlin; construction companies from Beirut; acoustic engineers from Eindhoven and Copenhagen; traditional architects from Jaipur & Venice; marble cutters from Agra; Television system consultants from London, and System integrators from Hollywood. Esteemed personalities and performers from Bolly / Molly wood Film & Television industries; musicians and instrumentalists from Chennai, Hyderabad & Thiruvananthapuram; stage & tribal dancers and other performers from across the country; artisans & craftsmen from all across the Asian continent joined forces with their expertise and skills. They consulted Zubin Metha, the celebrated conductor of several

Philharmonic Symphony orchestras around the world. The CAD drawings and 3D simulations /animation were done in Graphic design studios at Bangalore & Hyderabad.

As a special gift to the people of the state, the Sheikh wanted a versatile, exceptionally artistic, feature and technology-rich, futuristic Concert hall, superior to any existing anywhere in the world at this time. It's designed as a venue for all multimedia intensive enactments-music, orchestra, drama, TV & Movie, conventions, Teleconferencing, company presentations, speeches, and the performance arts. It shall provide an '*immersive*' effect for the audience & spectators. The dialogue, music, and the individual instruments have to be heard and located. The facial expressions & techniques of the performers have to be seen and understood in order to be enjoyed. So also the dance steps, the acts, and all the other motions performed on the stage must be clearly highlighted to be perceived. This can only happen with providing a detailed and magnified view by using visual technology, augmented (made bigger or supported) by the aural part of clear hearing.

In great performances, an intimate visual, aural, artistic sensory engagement between the audience and performers are essential. The actors need the assurances that their performance is clearly heard, understood, and seen at each and every seat in the hall, so that they don't have to '*outdo*' themselves compensating for any perceived deficiencies. The sound must be natural, balanced, clear, intelligible. It must have an appreciable resonance without causing echo, or howling noise; losing no consonants or vowels, without curtailing the pitch/ frequency range from *low* to *high*. The sound must also have good '*dynamics*' from very soft to the loudest range without distortion. All the design features are predicated upon achieving these main objectives, and stringent requirements.

The Sheikh and his family love the Monsoon rains; spends the whole of the season in the state of Kerala. Lovingly, he calls it his '*second home*'. He arrives well before the start of the season in early June for the prolonged South-West Monsoon lasting to September. He leaves for a couple of weeks thereafter; promptly back again for the short version of the other rainy season, namely the South-East Monsoon which starts in October for 5 – 6 weeks. Apparently, he watches the weather forecast very closely! The locals ask among themselves "when is the HRH (His Royal Highness) arriving?" instead of when does the Monsoon start this year! The Sheikh enjoys the deadening thunder, the intense lightning, though often become terrified by them. He had been sighted sitting outdoors, under heavy downpour, fully soaked until shivering grips his limbs and body. [Thunderbolts of lightning pack tremendous electrical power and water conducts electricity. So scamper back indoors!. ***Please inform the Sheikh!***] He says that he feels the greatest pleasure and thrill; "there's nothing more spectacular or enjoyable in life, and under the skies!". He loves the country; as a reward wants to be an enthusiastic recipient of all its natural bounty.

In the intervening period between the two rainy seasons, people hold a number of folk-festivals. They bring cheer and joy among them who have endured calamities of flooding, landslides, crop damage, and economic devastation. People view these misfortunes with an accustomed philosophy and traditional belief in supernatural powers giving them rewards or punishments. Accordingly, it seems that God is a strict punisher and stingy with forgiveness. No one wants to challenge the belief. The end result is that a few things that people can and must do, not get done at all!

Undaunted by the recently suffered severe setbacks, they look forward to the upcoming cultivating season with hope of better fortune, benevolent blessings, and prosperity. Only time can tell if they materialize or not. This is the cyclical fate and the perpetual destiny of farmers!

The Transformation of Manthravadi Mathappan

From an episode from Mathappan's Television show:

Often times, the fishermen of the South Western coast came back from their fishing expeditions in the outer seas with stories of dangerous sea monsters roaming and lurking there. They described how the creatures tried to pull their nets down and sink their boats. They were trying to devour and feast on humans. "They are the factual ogres! (the legendary man-eating monsters). It caused widespread panic throughout the fishing communities. People hesitated to come to the shore, even for buying fish cheaper. Their catch had been steadily decreasing, and demand dwindling. It was an unusual logic of market economics, the so-called theory '*supply and demand determines the price*'. People who would have been fishing at the time were frequently sighted loitering and scavenging on the beach. Their stories deeply and painfully affected Lord Parasuram who used to spend a great deal of time on the shores among them performing Pooja and '*Thapass*' for the continuous well-being of the population. He was determined to help the hapless community. Their survival and livelihood depended on daily fishing trips. He was deeply worried about them; gave thoughts to those stories. Immediately, he embarked himself on a series of rigorous meditations, contemplation, and planning; his magical "*Axe*" always beside him. He made figures, marks, and drawings in the sand; built sand models and replicas of exotic, and celestial objects. He was literally '*buried in the sand*' up to the neck for *days on end* (several days in a row) to gain concentration and to do penance in an earnest attempt to please the higher gods. He was sometimes seen pacing and measuring with his '*big foot*', up and down along the coast. He looked out into the vastness of the sea; returned; drew some more figures, and so on. As usual, the people left him alone; they never did comprehend what he was doing, or up to. Ultimately, Lord Parasuram was bestowed with certain divine divulgence. The heavens are many millions of light years far away. Messages do travel at '*light speed*' (300,000 km / second. One light year = $9*10^{12}$ kms!). Yet, it takes several years to go back and forth until the final message reaches the earthlings". Heavens must be very far away. People who are weary of travel may not get there at all!

Suddenly one day, the Lord, assembled an armada of fishing boats, all those were ever built and moored along the 850 km of the Malabar coastline; loaded each with bundles and rolls of ropes he could gather from far and wide. For months, people have been spinning coir yarn and weaving them into bundles for this purpose alone. Wherever he went, he carried his magical 'Axe'. On an astrologically important day, he joined the fishermen in their boats to the spot where the sea monsters were said to have been hiding in wait. The boats were spread across the entire span of the sea from the shore to this spot in spaces at 'rope lengths and calling distance'. This was the very first "Remote-Calling Network" in human history.

Up on reaching the spot on the leading boat, he tied all the ropes together end on end, extending all the way back up to the shore. The far end was wound around a humongous rock for leverage. Previously, he had arranged an army of strong men at the shore. From the first and farthest boat from the shore, he tied the loose end of the rope to his magical Axe and lowered it; probed, and foraged wide swaths of the bottom of the sea rowing around for a good length of time. [It's the same techniques the villagers utilize when they lose their lone water bucket in the bottom of the deep and narrow well that serves the entire community. As the well emits poisonous CO (Carbon monoxide) gas, no one ventures to go down and retrieve the bucket. They used a manual method of patiently probing, hooking, and lifting! Remember, the metallic hook has weight as well as certain magnetic properties!]. It took many attempts and hours of trying, until he felt that the Axe hooked on to something solid. Suddenly, he cannot extricate his magical Axe! It is 'inextricably entangled'! Certainly, it must be a massive, object fully submerged.

The fishermen shouted. "That's them, the horrible monsters, the '*Kelpie*! (a malevolent water spirit in Scottish legend, that lure travelers and cause drownings).

Now, you believe us? This is exactly what we had experienced!Let's quickly get back to our home shore, before the creature devours us all! They are trying to drown us; pull down the Lord deep into the water; soon, it'll have a nice tasty feast off him!". The fishermen crew imagined many nightmarish scenarios. But kept on staring at the spot in abject fear. At the same time, they had great regard, and faith in the

Lord's divine abilities. They had witnessed and experienced many of his prowess, might, magic, and strength. Unperturbed, knowing gods are on his side, the Lord gave commands to the people on the shore to pull on the rope, and pull it hard with all their might. After a long while and repeated trials, they felt some movement, albeit tiny little. The Lord's face was brightening up, indicating a subdued optimism!

The crew on the shore recruited more men, elephants, oxes, goats, dogs, spinning wheels and anything that has some strength, and use. It grew into a '*people cum animal*' fest. The drummers came without any invitation and beat their drums to pace the rhythm of pull. The pulling crew responded well with added energy and enthusiasm. It was, again a first case of '*joint venture*' of animals and humans in doing work. That was even before using bulls for tilling land or horses for pulling carts. They worked together in tandem, and parallel forces; complemented each other; supplemented the forces. People could immediately feel that they were pulling on some heavy object and trying to salvage it out of the water. But as it was coming higher and higher, became likewise heavier and heavier until it got totally stuck in the middle, somewhere between the depths and the surface of the ocean. It's now buoyed and balanced; the upward thrust of water pressure and the weight of the body balance out each other! It was a heavy and stubborn load; wouldn't move any further!

The lord was still optimistic! He looked around; saw some fishermen sitting in the boats, preoccupied with mending their nets with their primitive tools and needles. Immediately, he was reminded of sets of tools and materials brought with them that could be of some use. Among

them were sharp spades and poles; needles; bows and multipronged arrows. Many of the fishermen were skilled archers too, as they used the same technique for immobilizing large fish, saving damages to the large fishing net, and then scooping it out of the water. He used a combination of tools, and methods, and worked diligently. Eventually, the team manged to poke several holes on the stuck piece. Subsequently, due to the thrust of the water pressure acting from underneath and pushing upwards, together with the spinning action of the marine whirlpool systems, a few larger holes opened up; incrementally increasing in diameter, and finally into larger gaping holes. It functioned like a giant '*tunnel drill*' piercing through the mine! It was due to the combined forces of the tremendous water pressure and dynamic current at this colossal depths that the piece began to feel lighter and lighter as time went on.

He asked people in the boats to make additional holes. They did manage to make many more: 44 of them to be exact! Gradually, the water freely drained away through the drainage/ escape holes when the object was lifted. Slowly, this helped to lighten up the load. The rest of the task suddenly became manageable! After a series of rigorously long, and concerted efforts, the folks on the boats began to notice bubbles and jets shooting through the surface of the sea. The fishermen thought that they may have unwittingly angered the monsters. They are sneezing at them now in an ominous sign of extreme anger! The boatmen felt threatened. One by one, they began to fault the Lord who was feeling tremendous pressure from the fearful and non-believers. The utmost urgency did heavily weigh on him. Despite the fact, he managed to keep a quite calm and stern defiance. As time went by, the water around a large area was seen rising.

The first glimpse of the crust of the 'floated' piece of landmass with poked holes on them'

A very large object was spotted rising gradually to the surface of the sea. Indeed, it did! It was real! It came up floating, emerging to the surface with a violent splash, trenching all and capsizing many boats that were floating over the same spot. The boats were rocked; threw people overboard hurling them skywards and dropping them like pelted stones into the agitated seas. The rising crest of the waves also reached the same heights; caressed the thrown bodies on its crest; provided a form of 'foam-cushion', easing the pangs of the plunge. Spontaneously, the 'thrown men' contorted their body in the shape of the crest; used their backs as 'virtual surfboards'. After a series of 'roller coaster' ride, they landed smoothly gliding away with the waves. By and large, in the end, it was a soft landing in the water with only a gentle splash, as a parachute would drop the astronauts returning to earth! After a long while, when finally the waves and the turbulence subsided, they were all rescued without a casualty. It was an ecstatic and breathtaking occurrence for those who experienced the moments of aerial tumbling, and plunge. Nobody has hitherto reached such an altitude, defying gravity. In the same short period, each tasted fear, pain, anxiety, close encounter with a kind of 'end-of-life' scenario, weightlessness, fainting, ecstasy; ultimately, a satisfying relief. The gravity defying experience is so unique, it may not be repeated ever again!

Now, everybody was accounted for; a serene calm prevailed. Many new species of fish appeared swarming towards the '*new-found land*'. They were finding exotic and nutritious food on it, suggesting that the newly emerged piece was a very fertile matter. The fishermen's eyes glowed with glee and great prospects of easy catch. Instinctively, they threw their nets in the direction. [fishermen would never go out to the sea without their nets! That's their instinct, habit and unmistakable routine]. The Lord reprimanded them and reminded them that they hadn't come there this time for fishing, but only to meet the feared 'sea monster'. It turned out that the floating object was a huge landmass, larger than anyone had ever seen floating. It was not only floating, it was drifting with the wind and tide! The stuck piece is now extricated and liberated from the tyranny of gravity! It was actually the missing '*land*' left over from the '*cataclysmic rift*'! (A violent and sudden change in the earth's crust), the piece that was ripped out of the great African continent together with Madagascar. It remained stuck deep on the floor of the Indian Ocean. Parasuraman knew by his divine insight that the dynamic forces of swirling under-currents from the Indian Ocean and Arabian Sea, as well as the constant pulls of the physical landmasses of Africa, and Asia around it, have been pushing and nudging the submerged piece gradually towards the North-East. At this particular moment in time, its journey brought it to the closest Southwestern tip of India. It will soon reverse direction and drift back on its route, gradually away from this point. All these years he spent on the shore, Parasuram had been plotting its daily trajectory on the white sands on the shore, and heaping stones at strategic locations. He has been patiently waiting for this critical moment to arrive. This is the day when it finally happened!

In the uncontrollable euphoria at the amazing experience of the discovery, and the astonishing feat they had achieved, the fishermen raced their boats back to the shore competing to be the first one to make this *prophetic* announcement in great awe and embellishments to the anxiously waiting crowd at the shore. Each wanted to stake his claim and win the credit for its spotting if not the discovery and recovery. Parasuram and his boat was left behind alone, who was a poor rower, despite his many other superhuman, and divine skills. It took him several more days to reach the shore. He alighted from the boat and pulled on the landmass by his lone effort and might, as the rest of the crew collapsed on the shore drained of strength, energy, and in the overcoming of ecstasy; barely breathing! Parasuram left them alone, to recover in the course of time and *on their own strength*. But, he could not wait any longer. He pulled on the floating object; dragged it hard and fast with all his might fighting against the inertia, the constant pull of gravity, rip currents, and wind forces. Well, perhaps he was the one who invented the '*tug of war*'!

At moments, the physical and natural forces helped him at random, and on rare occasions. Gradually, the floating piece built up acceleration; gathered momentum aided by a timely storm; finally came rushing onto the shores like a speeding supersonic fright train. The people ran helter-skelter from the area in fear cursing the lord. With all the associated and natural forces of attraction between large bodies, the floating piece came crashing into the southernmost part of the Indian Peninsula and got fused into the Sahyadiri mountain range in a 'reverse process' of how it was originally split-off from the African continent! The assembled were harmlessly, and instantaneously scooped up and lifted onto the mount of the fluid mantel of the land in a big sweep, swipe and smooth ride to the top of the hill plateau of the Hi-ranges. This is the fastest speed ever achieved on earth, a millionth (10-6) of the speed of light! Suddenly, the life-long coastal dwellers found perched up on hill-tops, looking down to the all too familiar shore from a height and distance. They did certainly enjoy the swift ride, and the new view of the ocean. In a deep sense of nostalgia, they felt that the shores were beckoning them back, and the taste of fish profusely salivating mouths. They found it impossible to travel back the impassable trek at that much distance. Therefore, stayed there for the rest of their lives. They took up farming and cultivation for their living!

The continental amalgamation or land fusion caused the quickest event of mass migration of people in the whole of the human history. However, the resulting jolt knocked a loose eastern coast of the Indian subcontinent out into the 'Gulf of Mannar'. Well, the broke-away piece became the salubrious lotus island of Sri Lanka, a close neighbor on the Southeastern side of India! Thus, a maxim came into the everyday saying “you need to forsake something to get another thing!” (fishing is another example; 'bait to catch'!) It also proved Newton’s 3rd law of motion “to every action, there’s an equal and opposite reaction”! [I wish I heard this story while in school!]

As per the divine revelation given to Parasuram, the original piece of landmass, after splitting, did not have sufficient buoyancy to float and form itself as an island; soon submerged deep in to the bottom of the sea. For ages and centuries, it drifted back and forth, determined solely by the tides and the turbulent ocean undercurrents; waves, and other invisible but amazing marine phenomena. Nor did it ever find a land-body to attach itself to! All the forces it was hitherto subjected to, had overwhelmed its own. Few times, it was acted upon by two 'more or less' equal forces in opposite directions, giving it a temporary lift. As the forces became stronger, it elevated itself much higher above the sea bed. The fishermen were attracted to the resulting turbulence and currents, mistakenly taken it to be a large school of fish congregating for mating or spawning; or a humongous whale approaching. They ventured to the outer seas, deep outside their normal fishing comfort-zones. Instinctively, they cast their nets at the promising spot. The 'random theory of coincidence' occurred. Under those rare circumstances, a protruding part of this particular landmass was temporarily trapped in their nets. The unwitting fishermen thought they landed the heaviest catch in fishing history. They were not at all willing to let the 'bonanza' of lifetime simply slip away! If they catch that big fish, they don't have to go out into the dangerous seas for a long while. That was the bounty they constantly heard from their grandfather stories and were dreaming about. It were like 'Manna falling from heavens". They thought it finally came true for them!

The currents soon subsided. The piece of the trapped land began weighing the nets down; the fishermen will never let their nets go. It weighed down the fishing boats, as the net remained tangled around the object. When the brink of the boat came too close to sinking, they feared the worst; reluctantly, abandoned their precious nets and fled the area in fear and horror, rowing away as fast as they could all the way to the safety of the shore clinging to their spared lives. They blamed and cursed the 'monsters' for their torn or lost nets. Having reached home, they did not dare even to look back, thinking the hideous monsters will find out where they were living and come after them later.

The unsuspecting fishermen did not, at all comprehend what was really happening. Obviously, they could not peer far down into the opaque depths. They described their experiences to the folks ashore in their simple words and understanding. The fearsome stories spread, told & retold; metastasized, modified, and distorted by the 'word of mouth' re-tellings (in the study of communication, re-telling of stories invariably undergo individualistic, *stylistic transformation* known as "disagreeable personalization"). They were later interpreted and labeled as the fiercest "*sea monster stories*" of all times.

Only the Lord knew it better. He could clearly visualize the actual truth of the matter. He knew the severity of the task ahead. He needed more of the divine power and intervention than he already possessed. The gods themselves are generally stingy. They don't give away what is more than necessary. They want to maintain the power of '*dependence*' on them by the naive earthly folks! Well, that prompted Parasuraman to ask for additional help by embarking on a rigorous course of intense meditation. He performed several holy rituals. His heavy Axe had a lot of mass and magnetic properties. At first, he could not get any traction with the Axe onto the submerged land. He knew the *opposite bodies attract each other and like bodies repel*. So he polished and modified the polarity by reorienting the material particles in the opposite direction. [His multipronged Axe is the forerunner to the modern day '*anchor*' used to moor ships and vessels safely!]

In his infinite wisdom and reasoning, Parasuram had known many other related facts. He chose a day of astrological importance to perform this superhuman task. That was the day when a significant alignment of planets occurred when Mercury, Venus, Earth, Mars, in the order of distance from Sun and within the 'asteroid belt' were in perfect alignment in the 'Group of Foursome' or 'quartet'. At the same time, outside the asteroid belt, Jupiter, Saturn, Uranus & Neptune aligned parallel to the other 4 inner planets. Furthermore, this very unusually power formation took place on the same side of Sun! It's a unique and sublimely auspicious day! Those who attempt any task on such a day, are destined for success!

During the enduring course of time, violent ocean currents, and volcanic activities deep below the ocean bed had been frequently occurring. Under the impact of these natural forces, several pieces of the submerged landmass broke off in various sizes and numbers. Some drifted apart, some coagulated into bigger masses as they gradually cooled off following the cessation of volcanic activities. The forces of the whirlpool currents spun the pieces with violent centrifugal forces (the centrifugal force tends to move the objects away from the center when spun around an axis; the directional vector of the force is perpendicular to the axis. The same property is used in *Centrifuges* in isolating nuclear waste in research labs in highly controlled settings!) which ultimately hurled them away from each other, creating the archipelagos of the Comoros, Seychelles, Diego Garcia, Maldives, Lakshadives, and other Indian Ocean islands. At the same time, the *centripetal* forces (the forces tending to draw objects towards the center when spun around) acting towards the center, pulled many drifting pieces together making larger land bodies such as Mauritius, Andaman, Nicobar, and the Indonesian islands. They were deposited at ***geographically sustainable*** spots by the ocean currents at their current locations. They are not only surviving; they are prospering! There are marine forces similar to the celestial forces. In fact, all objects in the universe are constantly acted upon by many forces: push & pull; attraction & repulsion; tension & slack; gravitational, electromagnetic, static, and nuclear forces establishing balance and relative positions within it.

Many cults and religions make up similar stories, that have taken place out of sight away from the general-public surroundings. They authored their own legends and dished out to the gullible and the faithfulls. They branded the fabricated stories as divine epiphanies (illuminating realizations or discoveries) and cryptic messages addressed to the chosen few,- meaning themselves. The folks believed that the holy men possessed the wisdom and knowledge to understand and interpret them all; a fact meekly and readily accepted by the rest of us!

It is only because of the divine power, wisdom and intervention, the land now called Keralam has come into its existence; hence it earned its name "*God's own Country*!" It also remains as the only case of reclamation of a 'lost land' with human might and ingenuity! The lord returned to his heavenly abode, leaving the holes and scars as they were. The land is now left with a lot of them which people call 'rivers'! There are 44 of them crisscrossing the land, a region with the most dense with rivers, continuously irrigating and fertilizing the land in which they flow. People have soon forgotten that they were the originally poked holes by the costal fishermen captained by the lord Parasuraman!

Some researchers started examining comparative studies of the ecosystem, biodiversity, climate, species, geography, terrain, and other important factors of the two regions- Kerala & Madagascar. Some studies include the Ecuadorean archipelago Galápagos also in the mix. Many fascinating similarities now lend some credence to the story Manthravadi Mathappan has first narrated on one of his monologues in his Television shows. Initially, everyone thought that it was a hilarious joke. He got a lot of applause for it, ratings hit the roof. But religions branded it "*scurrilous blasphemy*". Consequently, he lost his job. But soon, he was recruited by foreign evangelical ministries and seen preaching at street corners attracting just a few curious passersby and lot less money in his pocket, drawing comments: "*Destiny is cruel*!"

In another twist of fate and events, not long after, with the additional insights and findings from Hubble telescope images, Mathappan's supposition has happened before, albeit rarely. In ancient cultures, wise men (Noticeably, wise women were branded as '*witches*' or '*sorceress*') knew about the celestial planets, its paths and trajectories at any given time. But for the fear of persecution by despots, tyrants, and religious fanatics, they were forced to keep the Science to themselves. In those days, anyone who propagates ideologies and theories contrary to the tenets of religion or established beliefs of the society was considered an '*heretic*', and pejorative (disparaging) of religious doctrines which demanded extreme punishment and imprisonment. In spite of it, wise men in particular concerns for the humanity, oftentimes, dared, but disguisedly applied their knowledge in certain imaginative manner, and clever ways. An analogous alignment of planets is expected to take place again in year 2022 when Mercury, Venus, moon, Mars, Jupiter, and Saturn line up in a '*remarkable formation*' with a significant difference of moon substituting mother earth in its position. With those scientific disclosures and evidences, everything changed for Mathappan. Lo, and behold! Mathappan once again became a celebrity for his astute imagination, sharp wit, and voluble skills of narration. He was invited back to resume hosting his show with a long term contract and additional amenities and bonus incentives! Immediately and smartly, he assumed a new stage name '*Angeles Mathappan*'. The church was tremendously pleased; granted him '*forgiveness*'; absolved him of all mortal sins of satanic worship. He is even permitted to practice '*exorcism*', considering the fact that he exorcized himself! (You know! The devils were once angels. They were thrown into hell for treason and jealousy. They must have secretly learned or inherited some powers during their brief stint in heaven albeit they use it for evil and vicious purposes).

According to an ecclesiastical promulgation of a decree, ceremoniously read out in Sunday sermon and worship:

"**From this time forward, Angeles Mathappan is hereby declared eligible again to participate in church services; bestowed upon him 'indulgences', and 'eternal salvation' rewarding his self-induced redemption, deliverance, and furthermore, virtuously embracing complete 'reformation' and 'true' beliefs. He is now a full member of the holy church. Forthwith, he may receive 'Holy Communion' and other sacraments as well, at will, and as necessary!**"

Signed: Your Beatitude and Eminence,

Cardinal Mar. Kuriakose Mathai Kadam-kunnathupuzha S.J.

More people began to tune into his shows. The sponsorship and advertisement revenue reached unprecedented levels. Up on gaining popularity and receiving the noble, blessed status, Angeles Mathappan was unanimously voted to be given the honor and privilege of 'MC', Master of Ceremonies to host the grand opening gala of the '*First Concert*' at the newly opening concert hall!

The hidden facts of Biological Sciences

The Sheikh has explored all the rivers on his inflatable dinghies watching bird sanctuaries and other scenic resorts; also in between catching a lot of fish. The dish of Karimeen & Varal cooked in coconut oil with hot pepper chillies and plenty of onions, '*Melam*' fish curry masala, lavishly added, are his other favorite things in life. He spends the harshest phase of the Monsoons, floating on spacious inflatable rubber dinghy built like a houseboat. He never anchors it; has no specific destination. It ends up in different locations along the waterways, drifted by the natural tide and ebb of the river. Maybe there's no river in his native country! There is a *rigged* sail that he seldom use. He lives by the rules of the nature, literally and figuratively!

During one of his monsoon trips, the Sheikh traveled from one end to the other of the entire 250 km long river *Periyar*, the longest and the most voluminous in the state of Kerala. He was curious to find out where it originates. He trekked through the remote forests of the '*Periyar Tiger Reserve in the Western Ghats*', and hiked the trails in the *Sahyadiri* (also in Western Ghats); explored the valleys, the puddles, earthen dams and mini streams. But failed to pinpoint its source. He wished if he could do the same for the entire 1500 Km stretch of the mountain spanning the states of Kerala, Tamil Nadu, Karnataka, and all the way up to the Deccan plateau situated in the state of Maharashtra. He gets distracted by many interesting and passionate things. He rues that there are too many of them and too little time left!

The region is a vitally important ecological preserve. Because of its biodiversity, origination of rivers and its prominent roles in influencing the weather patterns, it is often referred to as the '**Subcontinental Amazon**'. A very few of the rivers flow downhill towards the Bay of Bengal while the most flow westwards towards the Arabian Sea. The region is more of a forest than a jungle. The Sahyadiris provide a barrier between the Eastern and western coastal winds mitigating the effects of the weather and climatic winds affecting either side. The Sheikh has been making elaborate preparations to study the ecology of this region. He is organizing a team of trekkers. Included among them are surveyors, botanists & Zoologists, photographers & videographers.

One day, the Sheikh was fishing, moored at a remote part of the Periyar river. Suddenly, a boat is seen speeding towards his float. A much familiar reddish-brown, checkerboard-patterned '*Ghutra*' with white border is fluttering in the air from its flag post. A man in the traditional Arabic dress of white '*Thob*', red '*Ghutra*' and black '*Egal*' is seen standing on the mid-deck and waving frantically. He is merely pre-announcing that he is a friendly visitor and there is no need at all to panic!

The visitor calls out "*As- salaam 'alaikum!; As- salaam 'alaikum!*" *(Peace be upon you!")* . The Sheikh responds "*Wa-Alaikum-Salaam!*" *(And so unto you!)*
It is Akbar, his son. He is happy, but very surprised to see him appearing suddenly without giving any advance notice. He just wondered how Akbar did find out his location. *[Oh,... by location finder feature from his cell phone!. I forgot; he mused]*
"*Ahlan Wa- Sahlan!*" *(Welcome!)*
The father and son hug and kiss each other warmly on cheeks up on greeting.
The son places his right hand on his left leaning heart and makes a '*bow*' of respect and obedience.

"*Kaif halak?*" *(How are you?)*
"*Thamam, Al-Hamdu Lillah!*" *(fine, thanks Allah!)*, Akbar replies
"Wa Inta?" *(And you?)*
Quaise, El-hamdellah! *(fine, thank God!)*

They go back and forth with the same questions and responses a few times showing how wonderful to see each other; exchange other customary, rather lengthy rounds of Arabic niceties and pleasantries. Beverage refreshments of tea, sweets and light snacks are promptly served by the good attendant and trusted '*page*' (One who is employed to run errands, carry messages, or act as a guide in a hotel, theater, or club).

Then, they sit there speechless for few moments in quite introspection. Akbar seems agitated, sorrowful, and distraught. He could not control his emotions; let out a couple of suppressed sobs.
The father becomes very worried. He feared about his wife whom he left at home. Is she sick...gravely?...what might have happened to her...?.
All these thoughts rushed to his mind distressing him.
But he managed to put up a brave face.
Finally, the Sheikh mustered up some courage.

He cannot bear to see his favorite son so tormented like this.
"What's really bothering you, son?"
Words do not come out of Akbar, he chokes...
The father holds his son's hand in his, and give him a loving hug;
Tries to comfort him.

"I said the '*triple thalaq*'!" Akbar blurts out in confession.
"Nadia delivered her third girl!", trying to justify his decision.
"She is not capable of bearing a son for me!" he adds somewhat resentfully, and dejectedly.
The Sheikh was saddened and shocked!
He is petrified! Plunged in his seat in a brooding state, silent, staring at a frog carrying a load of its *frog-lets* on a half-submerged log nearby.

Suddenly, he got up; excused himself, asked his son to relax and wait right there for some time. Then he retired to his cabin room; was heard talking to someone on the phone in English for a long time, mainly discussing topics on science, biology, theory of probability etc.

After a prolonged conversation, the Sheikh returns; pulls his chair closer to his son.

"Listen son!... Sadly, we don't discuss many important topics surrounding life, mainly because of our culture, tradition, faith, and customs. Now I have to break away from all these constraints and prohibitions for an instant".

"Please listen carefully!. I have decided to be a responsible father, a mentor, and a teacher to my children. I need to fix my house and family before I can do anything for the public good.
I have to finally share with you what I have learned and felt, all along, but until now, was unable to do so. Just like you, I studied Biosciences at Princeton University. I presumed that you have studied the same things I have learned from there much earlier. So, I was complacent; kept quiet. But, now I see that's not quite the case.
I am going to tell you what exactly I learned and what I can still recall from my deficient '*memory files*' of my past education."

"Have you ever known that I was sent to represent our country and participate in the '*Human Genome Project*' (HGP), an international research effort to sequence and map all of the genes, known as the *genome*. It was about identifying all the approximately 20,500 genes in human DNA, and to determine the sequences of millions of chemical base of pairs that make up human DNA.
Do you, by any chance, recall any of these names, terms, terminologies, or substances?"

Akbar nodded in tacit affirmation.

"In our world and culture, they are all considered '*harraam*' (forbidden by religious faith). Do they not?
I never bothered to discuss with anyone ever; actually, wanted to forget as quickly as I learned for fear of getting a '*fatwa*'
I was quite horrified that I might get into troubles if I did not.
The studies and long nights of readings seemed all wasteful!" he lamented.

"This is how I may explain the Bioscience with some crude metaphors and analogies.

In our Baudouin '*village*' Sulaimania, there are '*families*'- Amari, Awab, Caleb, Chafika, Gaider, Thalal, Tawfik, and Umri. In all families there are members or '*individuals*'. They are -Adil, Asif, Aisha, Aida, Chiba, Caleb, Gamal, Gazbia, Fareed, Fatima, Talib, Tariq, Thaliba, Thamina, Usman, Umar, and many others. The four children from the Amari family marry from Thalal, Caleb, and Gaider, families, creating a family relationship '*A T C G*' bond. Similarly, four children from Awad family marry from Umri, Cham, Ganim families forming a family relationship 'A U C G' bond.
Then, certain families are rivals and at loggerheads, they don't seek any relationship between them.
In a village many combinations of relationship bonds are possible!

Let's say, we apply the same analogy to human cell structure. Then the human *Cell* represents the village; *Chromosomes*, the families, and *Genes* the individuals! The *DNA*, the particular Nucleic Acid carries combination of coded information of the 4 mentioned chemical basis (A, C, T, G) that instructs the RNA, the other type of Nucleic Acid, which is (A, U, C. G) chemical based to synthesize specific proteins. This is meant for the essential process of cell replication, related to growth and maintenance of life; also for a host of events taking place within our bodies, without our conscious knowledge, awareness, or intervention.
I am telling you all these things to bring you up to the following critical fact!"
The Sheikh continued his elaborate and in depth lecture, originally intended as a simple piece of advice.
"In humans, each cell contains 23 '*pairs*' of chromosomes, for a total of 46 individual helical strands. Most genes present in the Chromosomes are amazingly the same in all people, but a small number of genes, a miserly fewer than 1 percent of the total, are slightly different among people. How incredible is it!*Homologous* (Relating to genes that are derived from a common ancestor)

Twenty-two of these pairs, are called *autosomes*, as it looks the same in both males and females. The 23rd pair, the '*gonosomes*' (the pair of

chromosomes responsible for sex determination), as it determines the gender, is called the gender / sex chromosomes.
Again, as amazing as it may seem, it is the only pair that differs between males and females!
It makes all the difference! It makes our world, the humankind. It's just so...'o...incredible!

Females have two copies of the X chromosome. Only males have one Y chromosome obtained from father, besides also having one X chromosome received from mother. It's probably why the males always claim superiority, overbearing, and dominating. In the ensuing ego, and euphoria, they simply ignore the unique and vital roles the females play in human procreation, prosperity, continuance, and maintaining of life, starting with their very *unique gift of child bearing*. They take care of all the responsibilities of rearing, nourishing, nurturing, and caring; bringing up children, and the family; perform numerous other wonderful functions and duties in very noble ways only they can satisfactorily perform, and fulfill. Many thanks, salutes, kudos, and honor to the better gender!

Imagine! They do not only cook all the food we eat for our whole lifetime. They serve the food they lovingly cooked. They feed us! They shall not be disparaged, or marginalized in any way, son! That's what should be '*haraam*'!" He made his statement with great passion, firmness, and complete conviction.

"The '*mankind*', especially our own community, even today, is unreasonably harsh towards them. Women are, undoubtedly the most resourceful gender. The '*battle of the sexes*' and the discrimination are totally absurd and unnecessary. Since the beginning of time, the nature has clearly established and ingrained in our DNA the function and role of each of the gender so that we can sustain ourselves as '*humankind*'. It is inviolable; cannot be altered by any human force, campaign, court authority, legislation, decree, or thalaq!"

The Sheikh paused; looked at Akbar inquiringly.

Akbar seems puzzled. It was quite obvious that Akbar needed further explanation, clarifications; over and beyond, more reassurances.

So he continued.
"Well, to reiterate and restate in an easier way:
Each parent contributes one chromosome to each of the pairs, so that offspring get half of their chromosomes from their mother and half from their father in equal quantities. It's truly a '*joint venture of equal partnership*' at conception. Thereafter, the females play all the vital roles and contribute the maximum to the common cause!

- Females have two sex chromosomes of similar appearance called X chromosomes in (XX) combination.
- Males have one X chromosome and another, a much tinier chromosome, called a **Y** (XY) combination.
- Sperms have half the normal complement of chromosomes and contain either X or Y.

- Mothers contribute X chromosomes.
- Female embryos should have inherited an X chromosome from the father to form X X *genotype;* while male embryos must have inherited a Y chromosome from the father to form a XY genotype.

The presence or absence of the Y chromosome is critical because it contains the genes necessary to override the *biological default* of '*female development*', and cause the development of the male reproductive system.

To paraphrase, females typically have two of the same kind of sex chromosome (X X), and are called the *homogametic* sex [*gamete*: a reproductive cell (sperm in males, or eggs in females, having only half of a complete set of chromosomes] Males have two different kinds of sex chromosomes (XY), and are called the *heterogametic gonosome*. The sperm contains 22 chromosomes as well as either an X- or a Y-chromosome. A woman's egg cell contains 22 chromosomes plus an X-chromosome. The sperm that fertilizes the egg determines the baby's sex. Accordingly if an egg is fertilized by an X-bearing sperm, the resulting embryo will have two X chromosomes

(X X) and will produce a baby girl. If a Y-bearing sperm fertilizes the egg, the embryo will have XY chromosomes producing a baby boy. Since men's sperm carry both male and female chromosomes (Y & X) in *50:50* proportion and women's eggs carry only one female chromosome (only X, because her sex chromosomes are X X), the man's sperm holds the key to a baby's gender!

Well, now you see that only the father have the capacity to provide the Y chromosome that can result in the pregnancy of baby boy!
What does it tell you, my dear ?"
The Sheikh asked in probing and inquiring tone.
Patiently, he waited for Akbar to say something.

"Now I understand, dad!" Akbar responded brightly.
"If it all, it's anybody's fault, it is mine; unequivocally mine... I admit it... I do, ...indeed! No argument there,... at all! Akbar said decidedly.
I must now go back if possible tonight itself!" Akbar said feeling some guilt, repentance, shame, embarrassment, and haste. He gets up from his chair, pushes it off aside and hurries to get onto the boat.

The Sheikh holds him back and asks him to be patient, calm, and attentive for another little precious moment if he could.
"It doesn't mean, you are definitely at fault, nor you're less of a man. It's merely a case of '*incidental probability*'! as to which chromosome appears and more active at that critical moment of fertilization. Moments arc important in lifc, you know?; some morc significant than the others. They occur at random, beyond anyone's wish or control! That makes us human, my dear Akbar,...my son. That's the way it is!

The fact remains, that it is the male chromosomes that determine the difference in the sex of the offspring when fertilization occurs, as females have only 'X' to provide...period.

Also there's another important fact to bear in mind!
Genes determine the patterns of character or traits, health conditions, the very makeup of our '*being*', by the so-called *genetic code* which we all

inherit from our parents. If your children behave in certain way, or have certain health conditions it's actually a reflection of what the parents and forefathers really are like!

Of course, mutations of genes, the environments, and external factors may affect certain aspects, traits, and habits. Behavior and conduct of people are also strongly influenced by '*memes*' (an idea, a value, a pattern of behavior that is passed from one person to another through generations) by non-genetic means, as by imitation, traditions, or customs. *Meme* may be construed (an interpretation or translation) as the cultural counterpart of genes!

To summarize the big complex picture, for your quick understanding: the chromosomes hold the genes; genes, in them have both DNA & RNA. They are '*nucleotides*' with two different '*chemical bases*' in different combinations of **(A, C, T, G)** for DNA and **(A, U, C. G)** for RNA.

- DNA is the boss and gives specific instruction to RNA to manufacture certain proteins.
- Each gene has a sequence of nucleotides, which is specific for each gene.
- It may be argued that DNA & genes are one amongst the most important materials of life!"

The Sheikh utilized the opportunity to inculcate as much '*facts of life*' as possible in this 'once in a life-time *heart-to-heart*' conversation.

"I have been actually talking to you from my fading memory which I don't trust for complete accuracy. You must hear it from a very trusted and authoritative source. Therefore, before starting this conversation, I consulted with my esteemed former research colleague and renowned scientist in the field, Dr. Narasimha Swamy, PhD. He is the Dean of the faculty of Biosciences & Genetics at the University. He will be happy to meet with you tomorrow at his office.
I don't want you to do anything in a haste without correctly and fully informed.

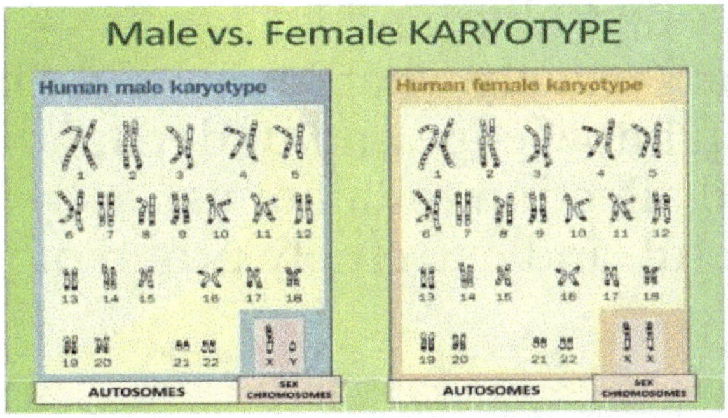

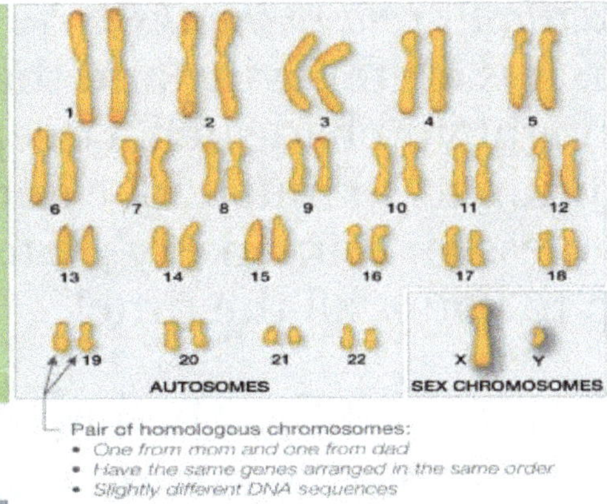

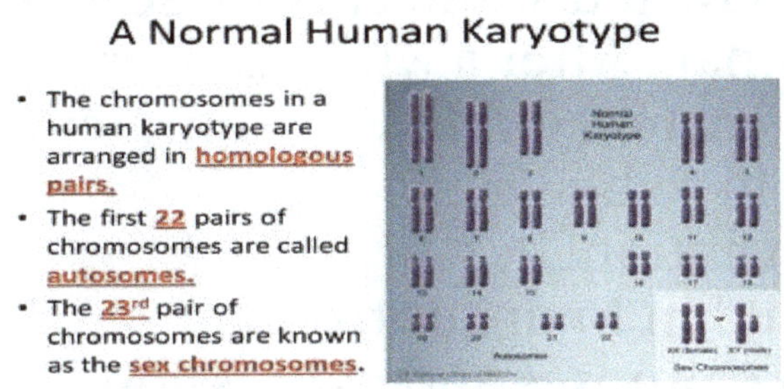

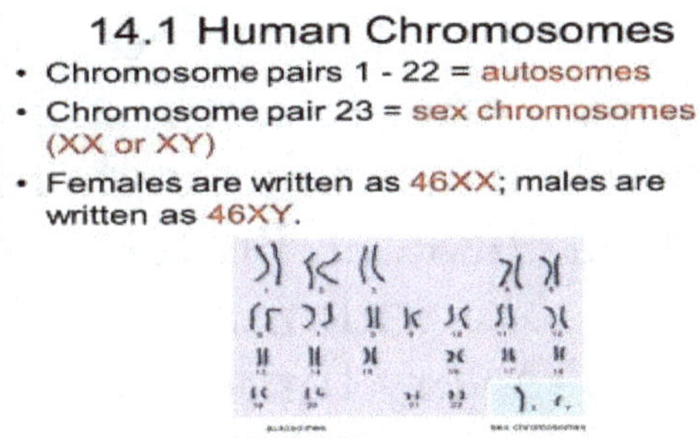

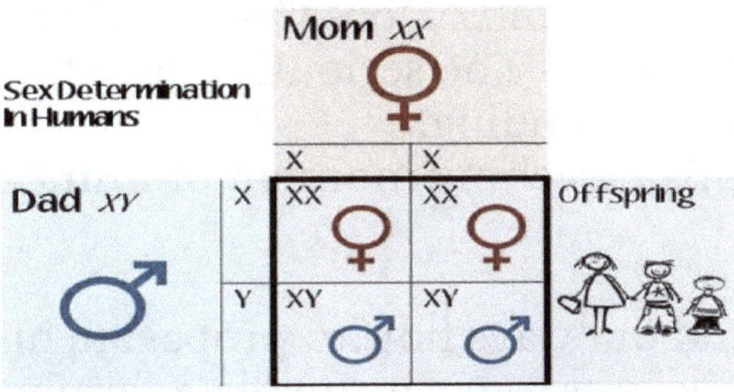

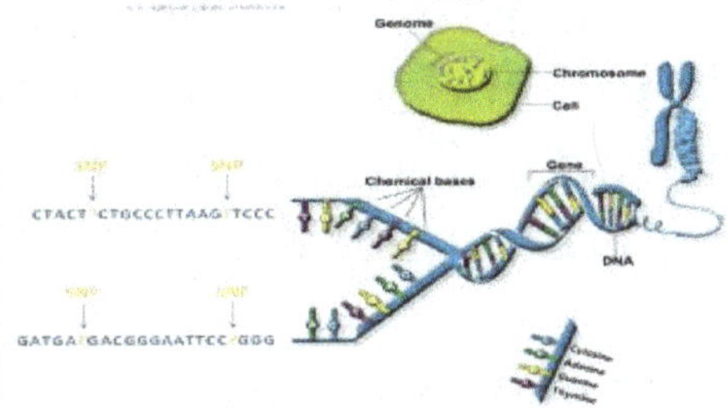

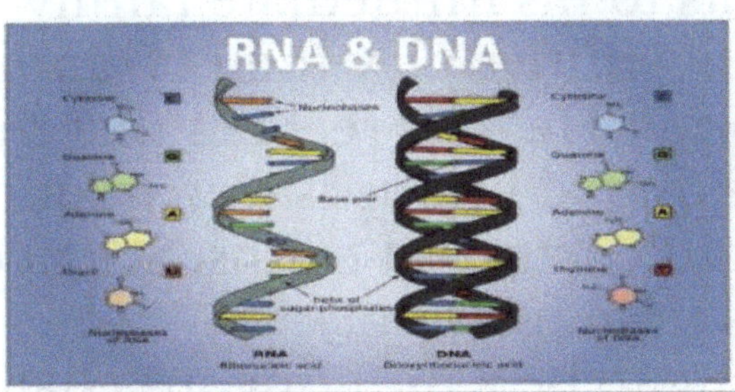

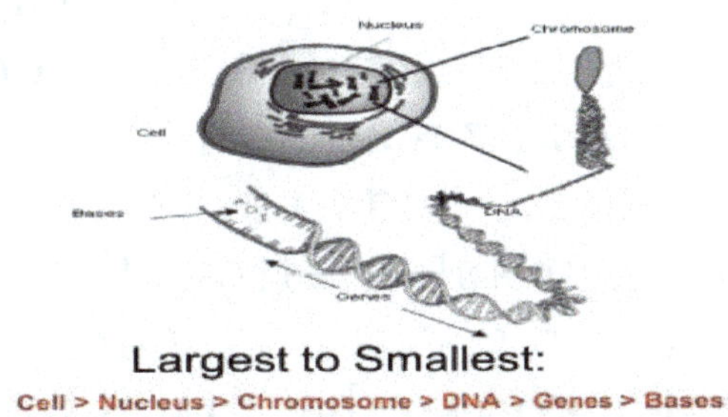

Purposefully, in '*Hadith*', (moral code of conduct / traditions / religious /sharia law) there is a provision for a '*grace period*' of up to 3 months to reconcile and nullify thalaq! I hope you know that! You have some time to consider all the relevant aspects. I shall accompany you for the appointment; introduce you to the scientist; then quietly step outside and simply wait until you are done, no matter how long that is going to be if that is OK with you!"

"*Shukran Jazilan*!" *(many thanks!).* Yes, dad I like and take your suggestion. It is quite loving and that's exactly what I need right now. I never thought you had such a deep knowledge and understanding on life; in addition to so much love and affection for all of us.
I am truly blessed; immensely grateful, and extremely proud of you; we look up to you,... all the time!"

"You know Akbar! I think Allah first created some chromosomes, paired them up; made man and woman. After a while....., Here we are! He didn't stop there. He mixed and matched different number of chromosomes. Other species came in to being.
All happened because of a single invention!"
....ha...ha...haaa...haaa....! .Both erupted in to uncontrollable laughter, qualifying the joke with many Arabic phrases and expressions.
"Baba (dad), You're very funny too!!...oh Upi!.... Does Omi (mom) knows about it?. You're great.....one of a kind!...." The seriously started conversation ended in a light-hearted ending.
It became an enlightening session; a teaching moment of immense value and consequences.

Akbar is now obsessed to find out ways how to propagate his newly acquired knowledge and insights to his immediate family. He wondered how traditional and external entities play their tyrannical roles in people's lives and on their minds. He now realizes that many natural facts presented to us from childhood onwards were totally distorted interpretations from the actual facts, and truths. Science has the most favored status in my life this time forward!
Of Course, I acknowledge that it's going to be a humongous challenge to clearly discern between blasphemy, Science, facts, and at the same time, to live by conscience, as well as the principles of truth!

Computer Analogy of DNA Structure:

DNA resides in the **nucleus** of cell.
The DNA is **tightly packed**
The tightly packed form of DNA is **Chromosome**
Cell Division
Complex organisms have cell structures called **"mitochondria"** containing small amount of DNA**,** in **egg cells,** Besides the common **Necleus DNA**

A **folder** in a computer Hard disk;
The folder contains **compressed** files
The **compressed file**
Copying the file
File may contain different forms of **data**

Mitochondria generate the **energy** the cell needs to function properly.

system battery

In sexual reproduction, organisms inherit **half of their nuclear DNA** from the male parent; **half of the nuclear DNA** from the female parent.
The offspring inherits Mitochondrial DNA only from the female parent.
This occurs because only egg cells, and not sperm cells, keep their mitochondria ***during fertilization.***

*An organism's **complete set of nuclear DNA** is called its **genome**.* ***File properties.***

DNA** is made of **chemical building blocks** called **nucleotides.
Nucleotides contain:

1. ***a phosphate group,***
2. ***a sugar group***
3. ***one** of **four** types of **nitrogen bases***

*To form a strand of DNA, nucleotides are **linked into chains**, with the **phosphate** and **sugar groups alternating**.*
The four types of nitrogen bases found in nucleotides are:

1. ***adenine** (A),*
2. ***thymine** (T),*
3. ***guanine** (G)*
4. ***cytosine** (C)*

The order, or sequence, of these bases determines what biological instructions are contained in a strand of DNA. For example, the sequence ATCGTT might instruct for blue eyes, while ATCGCT might instruct for brown.

*The complete DNA instruction book, or genome, for a human contains **about 3 billion bases and about 20,000 genes on 23 pairs of chromosomes.***

*The double helix structure of DNA enables it to carry **biological information** from one generation to the next.*

Functions of DNA :

Produce proteins, which are the complex molecules that do most of the work in our bodies.
DNA contains the instructions needed for an organism to develop, survive and reproduce.
To carry out these functions, DNA sequences must be converted into messages that can be used to produce proteins, which are the complex molecules that do most of the work in our bodies.

Each DNA sequence that contains instructions to make a protein is known as a gene. The size of a gene may vary greatly, ranging from about 1,000 bases to 1 million bases in humans.
Genes only make up about 1 percent of the DNA sequence.
DNA sequences outside this 1 percent are involved in regulating when, how and how much of a protein is made.
Different DNA sequences make different proteins
DNA's instructions are used to make proteins in a two-step process.

1. Read and Transcribe the information in a DNA molecule:

First, enzymes read the information in a DNA molecule and transcribe it into an intermediary molecule called *messenger Ribonucleic Acid*, or mRNA for short.

2. Message Translation

Next, the information contained in the mRNA molecule is translated into the "*language*" of amino acids, which are the building blocks of proteins (production order). This language tells the cell's protein-making machinery the precise order in which to link the amino acids to produce a specific protein.
To make a wide variety of proteins from the 20 types of amino acids, by mixing and adding required amounts in many possible combinations

The Amphitheater is a large, grand circular structure having tiers of seats, rising gradually outwards from the large stage. It looks imposing; well suited for its designed purposes. On the outside, it looks like wearing a massive, proportionally shaped, tipped '*top hat*', the dome. The '*brim*' of the hat is aesthetically curved; the facade, (the main front face and entrance of the building) is accentuated by a sculptured healthy bunch of golden coconuts; and at the balcony end, a heap of fruits- mangoes, banana, plantain, pineapple, papaya, pomegranate, jack fruit, cashew, '*Betel leaves/ areca nut /tobacco bundle*'; and an assortment of sweet & sour berries. They are made by chiseling on shiny marble, or carvings on reflective and colored ceramics, fashioned in modeling clay, and cast in metals of the true colors of the produce. Embedded beads of LED lights highlight each piece in its visible shape and form.

Inside, extending into both directions, stay equally spaced ornate Corinthian columns like those of the Mughal architecture with its capital (platform at the apex) decorated with a full bunch of young golden "*Pookkula*" (coconut flower buds). On them, are life-like mosaic statues of eminent figures of the ancient, as well as from the recent past; Ashoka, the emperor; Valmiki, the author of Ramayana; Veda Vyasa, author of the Mahabharata; Gautama Buddha, the enlightened; HH. Marthanda Varma, Maharaja of Travancore, who defeated the Dutch colonialists; HH. Sakthan Thampuran, Maharajah of Kochi; thc Malayalam poet, Kunjan Nambiar; Karnatic and Hindustani music composers, Kalidasa, poet and playwright who wrote many works and compiled the '*Kamasutra*'; Arjuna, the ace archer; Vivekananda, a teacher and exponent of 'VEDANTA PHILOSOPHY'; Raja Ravi Varma, the celebrated Keralite painter and artist, Aryabhatta, the astronomer mathematician, Ramanujan, the math prodigy; most famous writer Premchand in Hindi literature, and the inimitable and peerless contemporaries of the music, the '*Trinity*' of the 18th century Karnatic music- Thyagaraja, Muthuswami Dikshitar, and Shyama Shastrikal; Allauddin Khan, Pandit Ravi Shankar & Mohammed Rafi.

Some statues are carved in stone, housed in architectural stone cuttings. Others are wall mounted. The renowned architect who designed and erected the "*statue of unity*" at Sardar Sarovar Dam, Narmada, the tallest in the world was the chief consultant in this project too. The icons and '*figure forms*' of gods are sculptured in glistening bright bronze.

One of the wonders of the world, the magnificent white-marbled mausoleum Taj Mahal

The court-yard follows the same rectilinear shape for the long pathways to a flight of entrance-steps. The attendees are led to the large entrance to the porch built under a sprawling '*terrace*' (A platform extending outdoors from a floor of a house or building) supported by porch pillars bearing the '*axial load*'. The terracotta colored broad pathway is divided into two separate lanes on either side of a shallow, artificial, reflecting pool, with slender-trimmed & manicured flowering exotic shrubs planted on the curb-sides. It has striking resemblance to the courtyard of Taj Mahal in Agra.

The archway, foyer, vestibule, and colonnade-lined passage way leading to the concert hall

The edifice is built in Mughal architectural concepts of roomy rectilinear partitions, perfectly vertical symmetrical columns, archway entrance, and large domed roof. A posh portico connects the exterior to the interior of the '*foyer*' (a large open area just inside the entrance where people can wait and meet others; also where coats and hats are kept during the event). The '***marquee***' (a roof like structure that projects out over the entrance of a theater to display sign such as the name of performer, or performing troupe) displays '*Today's event*' in flashing calligraphic writings.

- ## The Balcony

The curvilinear balcony has an upper, middle, and a lower sloping gallery tiers giving spectators a high, moderate, and lower profile close-up views of the performance on the stage. The fine details of performance on the stage and the orchestra pit can be clearly viewed from every seat without leaning, straining the neck, or obstructing others' view. The roof and the dome are located quite high, leaving ample '*headroom*'. There is no feeling of congestion at any point. The 66 columns supporting the tiers are adorned with replicas of archaeological artifacts from the 3000 BCE Indus Valley civilization that flourished in the greater India consisting of present republics of India, Pakistan and NEastern territories of Afghanistan. There is a total of 198 of them in 3 levels, geometrically spaced around the circumference. Soothing cool air, enriched with pure Oxygen (a by product from the green Hydrogen plant situated nearby) is let-in through the carefully placed vents. The amount of moisture is closely controlled for each zone. They are also for the purpose of aiding the propagation of sound, and naturally enhancing pitch resonance. The powerful '*draft fans*' and blowers operate far away from the structure circulating the air causing no concern for noise, turbulence, or annoyance. The environmental parameters are optimally controlled by panels, devices & systems from Johnson Controls.

An elegant sky-lit atrium (a large open area in a public building, often providing a large view from there) leads to a large vestibule, a spacious lobby entrance or antechamber. Three elevators, each destined for a designated level, can take the attendees to the balconies. The surrounding peripheral vestibules on the ground floor lead to several entrances into the inside of the theater. The guests are taken on a horizontal escalator stopping at each entranceway. Prompted by the "*touchless*' face recognition or ticket /QR code scanners, the attendees are ushered to their respective seats by "smart ushers" by greeting each by name and directing them with visual, graphic, and aural guidance.

The magnificent Concert Hall

a general view of the multimedia-ready concert hall under construction

In the Indian culture, music and dancing are both divine; they are interrelated; the hymns are repeatedly recited in chants at religious rituals; pleasing the gods; beseeching blessings. The gods themselves play musical instruments and perform dances. In those days, the singers were mostly men while the dancers were females. Lately, the audience began to relish the beauty, richness, and sweetness of the high-pitched female voice. They petitioned and clamored for the inclusion of female voice in all concerts. Gradually, it became the better part of the musical performance!

The '*Gandharva Veda*' is specifically the science of music, dance, and poetry which was formulated in ancient times (believed to be written in 1000 BCE). Vedas are very descriptive treatise of structure, style, aesthetics, technical, and performance related rules and guidelines. [This is in contrast to the Western methods of using formulas, equations, or acronyms (an abbreviation consisting of the first letters of each keyword in the name of something, and pronounced as a word)]. Each dancing pose has a specific name and meaning. The Vedic literature describes musical instruments and singings to be performed during religious ceremonies. In its long development, the pundits recited; monitored what was said; how the listeners responded to what they heard. They debated until consensus prevailed among them, or proof obtained. It must have been a long process like the legislative bodies enact laws in modern democracies. Vedas must be categorical, undisputable and fail-proof! On the other hand, the legislations are '*politically motivated and intended*'; ambiguous and ambivalent enough that the judges can interpret them according to their political ideologies.

The tediously compiled knowledge, structure, and compositions were irrecoverably lost through historical events, upheavals, and in the course of passing of time. Now, it requires many suppositions, lengthy discussions, complex interpretations, and intricate restoration efforts to find the true and original forms. In the contemporary times, unwittingly, these efforts often lead to an entirely new system or a more human intensive music, rather than the original divinely focussed structure. In Karnatic South Indian music, the traditional influence can be felt as still tenaciously persisting. In the present day, the essence of music is mainly to celebrate harmony of sounds with the nature, and appeal to human emotions; create coherence, and consciousness of mind in simple thoughts and perceptions. Naturally, this, in turn, brings forth mental balance; positive vibes, & temperament; importantly, relaxation & healing to the singers and listeners. The faculties of mind, imagination, and senses sharpen and intensify; turn positive, productive, and refined.

A closely related subject is the '*Natyasastra*', (Sastra meaning Science) which is the science of dancing and choreography. It explicates the intimate relationship of music and dancing, especially when it comes to the rhythmic structure, accent, tempo, dynamics, emotions, etc. and how well each of these elements may be best articulated, and performed. The music gives relevance, meaning, emotion, and appreciation to the dancing actions and poses, without which the dances remain merely a '*mimicry*'; some meaningless actions, or motions. In certain symbiotic manner, dance provides the music another tool in the interpretation of its tune and lyrics due to the fact both contain common elements. We could say dance is a combination of *portean* (readily taking on various shapes or forms, as music combines vocal & instrument sounds in rhythm, melody & harmony) art form and performing art form. Both are expressive and appeal to the same, and sometimes identical faculties of human senses. Beyond the theme of dance, Natyashastra emphasizes interrelationships; offers suggestions of several aspects of vocal, instrumental, as well as orchestral compositions for both components of the art forms.

The facial expressions in Indian dances provide deep and meaningful impressions, and help to tell the story more effectively. Besides the hand, body, and step coordination, the delectable and changing group formations (choreography), as well as the energy displayed in dancing add elegance, and aesthetics to the over all performance. The dance poses are hugely important too. The dancers wear foot bells (*chilanka /padasaram*); the clangs produced by the dancing steps synchronize and coincide with the music played with all its finesse, flair, and mastery! It's the '*aural effect*' and hearing part of dancing in addition to its visual impressions. The sound may convey specific mood and emotions, such as joy, wonder, surprise; hesitations, suspense, haste; acceptance /agreement, rejection / disagreement; sorrow, or ecstasy, and so on. It touches and forms the listeners' /viewers' emotions! The feedback is essential for performers, as this is the main measure of the quality of their repertoire. This is similar to the cheers or jeers the sports people receive, often affecting the outcome of matches. The Western dance routines tend to focus on the elegant strides, postures, gymnastic, athletic, sporty movements, and body flexibilities; involve hoping, jumping, spinning, gyrating physical activities with vigor and intensity. These physical acts require much effort and energy. The strain caused by it usually become evident on the faces, and also the muscle tension on the limbs.

Chapter: 9 The Wall paintings

The paintings on the theater walls are meant to evoke and refresh memories of the legacy of the Vedic Sciences of fine & performing arts from the ancient times and connect with those of contemporary times. They tell a lot of tales! The theater walls are adorned with paintings, and delicately embossed carvings, in the available space above the implanted Television and multimedia gadgets. They are thoughtfully chosen from the vast array of Indian musical instruments, art forms, images of the gods who inspired them; the artist's depiction of *Gandharva Sangeetham, Kacheri* (the open-air concerts); the heroes in Indian culture & mythology; faces of ordinary men and women of each of the states and regions across the country, their arts, and costumes; the festivals of India; significant Sanskrit verses from the Vedas, Upanishads, Puranas (Legends) and the Epics; the portrait of meditating Shri. Buddha's enlightenment; his teachings; historical events; wars & freedom struggles against the colonialists; the plundering of the much coveted jewels and wealth; the persecution, inquisitions, impositions, and taxation under the British Raj; the portraits of great men and women, symbols of religious tolerance and harmony are prominently and purposefully depicted.

Besides being works of art, these paintings are evocative visual learning tools of many subjects. The creative and non-verbal communication enhance imagination, intelligence, and render several social benefits. It promotes social interactions. The usefulness of images in multimedia communication is evident in every walk of life: in education, presentations, design, simulation, modeling, etc. Drawing, Painting, and Arts are measurements of values, culture, moods, and powerful expressions of a society. The artists themselves have developed good motor skills, memory visualization, imagination, creativity, and coordination. They are especially thankful for the opportunity to utilize their talents, expand their horizon, and earn a living. A talent which is not routinely used, meets with a slow death. A work of art is likely to have longevity longer than its creator. Its influence and effects are far-reaching, as clearly seen here and confirming the adage "a picture is worth a thousand words!"

Murals depicting life, culture, history, and festivals

The Sanskrit poetry, being of divine origin, reigns supreme in all the literary works of mankind, dating back to the ancient writings. It follows stringent codes of structures, rhymes and rhythms,

meticulously measured in vowels and consonants, presented in profound meanings and exquisite expressions.
Understandably, *it is not spoken, only reverentially recited in poetic style and lilting cadence !*

Poems in any culture or language have certain characteristics and common structure. In addition, they have a unique form and feature related to the afore-mentioned factors. The stanzas in poems are comprised of lines of verses mostly of specific number of lines [Couplets of 2 lines; Tercets of 3 lines; Quatrain of 4 lines; Cinquain of 5 lines; Sestet of 6 lines; Septet of 7 lines; Octave of 8 lines, etc.]. They are so formed to present a specific idea, statement, or sentiment using rhymes and rhythm of cyclic sound pattern as it is read or recited. In order to sound like a poem resembling an '*art song*' (a lyric song intended to be sung in recital) the verses /lines in a stanza must conform to certain rules of '*meter*' which means they have to have the equal number of syllables, in the lines of a stanza, some '*heavy /stressed*' others 'u*nstressed /light*' in strategical places. This tells a poem apart from a '*prose*'. Similar sounding words (rhymes) in identical positions in each line make the recital (tone) of the poem pleasing and an interesting read. Poetry uses '*figurative language*' with metaphor, conjectures, similes cultivating imagination, impression, and usually better understanding. '*Prosody*' describing the set of speech variables, including rhythm, speed, pitch, and relative emphasis, that distinguish vocal patterns, is an important element of poetry. So also the '*cadence*' which refers to the balanced, rhythmic flow with inflection or modulation of the voice make the recital of the poem captivating. Some genres adhere to a typical style. Sanskrit poetry is one such genre!

The repertoire of classical Sanskrit poetry uses mostly Quatrains (also made famous by Nostradamus by writing his entire book "Les Prophéties" in quatrains!) using either specific number of syllables, or '*syllabic time-lengths*', modeled after the Epics of *Ramayana*, and *MahaBharatha*. That was the accepted literary style of the time! Strictly speaking, all Vedas are highly prescriptive; Sanskrit poetry is not any different. The number of syllables, the types of syllables, and their '*groupings*' are rigorously followed throughout the work. As an example, the heavily accented syllables may be suitable to refer to undisputable and divine beliefs, while unaccented ones may be for common, or mundane suppositions. From the ancient times until relatively the recent times, all important knowledge and teachings were delivered and pronounced in the '*all too familiar*' poetic style. Therefore, it plays a very significant role in the culture, education, conversation, music, and arts of India. In deep respect of these considerations, the Vedic scriptures are inscribed on the side of each colonnade in and around the concert hall. It takes many columns to accommodate excerpts from a single work. The parts of the same subject matter is identified either in color, or a Vedic symbol. All the 72 "parent" ragas (Raga is a framework of notes compiled in specific patterns expressing certain musical qualities, tonality and motifs reflecting Indian traditions and thoughts) are depicted on walls in its ascending and descending order. The same goes for the "Talas". (Tala is a rhythmic structure comprising of beats in a time meter/ measure. It may or may not be repetitive in the same manner in every measure)

A collage of wall paintings

There are numerous fresco art & mural paintings. They represent '*folks arts*' showing paintings of rural life and struggles of survivals, carrying burdens and bundles; mending cattle and sheep; tilling fields with two oxen bound under heavy yoke; women tending plants and flower gardens; picking tea leaves in the plantations in the High-ranges of Munnar, Peerumedu, Vynad, Assam, and Darjeeling; collecting coffee beans in to the '*back-pack baskets*' strung behind the back; hand plucking cardamoms and cloves; apple and lemon harvesting by beautiful Kashmiri damsels wearing fashionable head-dress. The village maidens carry water in the '*pyramid of pots*', stacked one on top of the other, yet another pair, one on each side, tucked at their curvy hips;

balancing the burden with amazing ease, and skill. They chat merrily and dance while walking with their '*head-loads*' all the long way homes along the unpaved paths made solely by constant walking. The circus acrobats, jugglers, and magicians perform their skills, entertain the gathered at street-corners for some applause and a few pretty pennies. The bystanders look away, casting a flirting look at the parade of the '*pots-bearing*' beauties while they pass by in colorful parade. They blush and show a sly smile to themselves in some pride and self-recognition. At another place, people worship in many faiths and customary rituals; children play games, fly kites, and ride bulls; the shepherds collect the flock, blowing into Conch shell; herd them with a long cane which has triple purposes as walking support, and also as a low-pitched flute the sheep always enjoy hearing.

An assortment of mural paintings on the walls

There is an allegorical (representation of abstract ideas or principles by characters, figures, or events in narrative, dramatic, or pictorial form. Today's cartoon series do the same) painting that re-enact the "*Lanka Dahanam*" (the burning of Lanka) in the epic war, described in mythology. The '*Vanara Sainyam*' (the irrepressible army of Apes) lead by its '*commander-in chief*', Lord Hanuman, bridges the '*breach*' (an opening) of the *Palk Straits* hanging on each other's tail. The water flowing beneath, reflects the raging conflagration all the way across the passage way. The *Pushpaka Vimana* *(the flying chariot of flower bed, piloted by divine powers)* flies across the straits between India and Sri Lanka with its retinue of soldiers to rescue the kidnapped wife of Rama, the beautiful princess Sitha. The events are colorfully depicted in a hanging mural spanning across the span of two columns.

Then, there are paintings of weather and natural phenomena in its fury and calm; the rains & the flood; the flora, fauna & fountains: landscape with hills, cliffs, and valleys, rolling ridges; cascades of steep waterfalls; trees & palms, rustic meadows; plantations and groves; rivers, the beautiful bungalows on their banks plastered with pebbles & granite; the decoratively painted tattoos on the hands & palms of the Indian brides; the clanging glass jewelries worn by the poor village maidens; the tribal dancing & tribal arts; the brides and wedding scenes of different faiths and religions; smiling faces of innocent villagers; the grave expressions of business people; the wickedness of the scoundrels; the waste, excess, luxury, lavishness, and splendor of the super rich; the snobbishness of the élite; the faces of '*downtrodden*' men & women with their scars of daily toils & travails, wrinkles of age, wisdom lines, frown lines & sweat valleys; the statues reminding of the piety of the deities; so on and so forth are depicted with utmost sensibility and impressive artistic originality.

Elephants obey commands and carryout tasks; make cranes redundant and unnecessary

The elephant, also the state symbol is very dear to Keralites. It is held in respectfully elevated status than that of horses in Western countries. The elephants play important roles in temple festivals, logging operations in the lush, rugged tree forests on the Eastern hills and Hi-ranges. They are the '*living cranes*' who are fast learners; obediently follow commands; has great intelligence, diligence, and reflexes inherent in them; are regularly used in building projects. They are rewarded with a sumptuous feast of green leaves on supple branches and plenty of water to drink; pampered with a cozy, massive massages, and bath rituals. The giant, but kind animal has also served in many significant ways in wars, mainly as transport vehicles. They played deciding roles in the wars of *Jhansi* and *Panipat* against the British colonialist aggressors, and occupiers. The elephants can also be vindictive, (*disposed to seek revenge; revengeful*) if abused or insulted!

Elephants, wild, domesticated, workmate & co-passenger; processional leaders; baby elephant as a cute pet; friendly, and even religious

The elephant herds are credited as the first ones to detect the *X'mas day 2004 South Asian Tsunami* as it happened, much ahead of all the sonar, satellite based '*sophisticated*' systems, saving many lives of humans and animal species. Their big feet firmly planted on the ground to support and balance its massive mass, felt the tremors seconds before the onset of the Tsunami, trumpeted to the herds and lead them to the relative safety of the higher grounds. Although, the satellites had spotted the phenomenon, it is said to have taken 5 lengthy hours to process the images and transmit the message to the world.

Schade! (Ger: it's a pity); much too late to prevent tragedies!
The elephants are social animals!. They are first class ***socialists***!
They don't have any traits of the selfish, arrogant, and wasteful capitalists.

Humans owe a lot to the gentle giant of the forest, who also plays a gigantic role in preserving their habitat and people's lives! Honoring its versatile contributions, the Elephant is prominently shown in its various roles in many places on the canvass though it takes up a lot of space! It is significant to mention here that there is no report of poaching for ivory trade in the state, quietly serving an exemplary lesson to the notorious ivory trading port of Zanzibar.

The pieces of the artworks are carefully laid out in a meaningful order and groupings on the vast real estate of the theater walls in pleasing, evocative, awe-inspiring reminders. Eminently knowledgeable and skilled artisans painstakingly worked for several years to accomplish this work of art. Many local media sources compare it to the mural paintings on the dome of *St. Peter's Basilica*, in Vatican City, the papal enclave located within the city of Rome. Significantly, the foreign media critiques made no such comparison!

The huge stage is endowed with innovative design and features. It differs from any other conventional stage. It has a flexible background wall, meaning the same wall may be altered into a curved or flat shape depending on the requirements. A video wall may be pushed along the '*recessed*' (set below surface) rails and set in place of the wall. Similarly, all the movable items of props and fittings can be slid along the '*grooved*' rail systems. They may have a matrix array of UHD flat panel displays fitted with an acoustical shell.

The stage is divided into two major areas, a performing stage and a '*backstage*'. The performing stage is built on curved individual sections. They are assembled in abutting (contacting the adjacent piece) sections. Some selected sections in the rear are mounted on heavy duty hydraulic columns. They can be individually raised if needed by incremental levels from front to back giving each of the four sections different elevations. This is primarily for the use in orchestral and choral performances. The stage can easily accommodate a 100 piece ensemble, plus a choir standing on 4 level '*choral riser*' with its own guard rails fashioned like the hospital beds. When a large stage choir is performing, it becomes the main attraction. Then, the orchestra is *relegated* to the orchestra pit. With this horizontal-vertical arrangement, not only that all the instruments and players can be clearly seen, but also they appear closer to the audience.

On a conventional stage, all or most instrumentalists are on the same level. This arrangement, obviously obscures many performers and their instruments. The situation is exacerbated by the big piano; likewise the conductor on the '*conductor's podium*' showing his rear to the listeners / viewers. Additional problem is caused by the big note stand and the Conducting Attire worn by the conductor, a tuxedo, or a long tail coat. Conducting is the art of directing the simultaneous performance of several players or singers by the use of a code or sign language of gestures. The actions of a conductor precede the flow of music score indicating how the upcoming '*measure*' (a specific unit of time defined by a given number of beats) is to be renditioned. The job of the conductor is to bring a piece of music to life and to its emotional heights using various left-hand gestures, facial and physical expressions, while the ***baton*** in the right-hand usually dictates the tempo, accents, and delicate musical movements. A downward stroke of the baton signifies an upcoming '*downbeat*' indicates the '*first beat of a bar /measure*'; whereas the '*upward*' stroke alerts the '*unaccented beat*' or the last beat in a bar and the '*upcoming*' of the next measure. The '*heaviness*' of the movements (subtle, or heavy) indicate the amount of the accent to be applied. The left-hand prompts the musicians to enter or exit the music as the case may be, at specific times by identifying/ pointing to an individual musician /instrumentalist, or an instrument group (like the maneuvers of good old traffic police). For '*canceling out*' a solo or group participation, the conductors folds his fist; looks away and /or towards the others. The speed, intensity and other nuances in hand movements signify the specific manner a music passage must be played. An open palm holding steady parallel to the stage floor asks: "maintain '*status quo*' in terms of speed, dynamics, and style as it is". When a *crescendo* is needed, the conductor has to make two simultaneous signs; one showing the level or intensity; the other to indicate the pace or speed of execution. He/she *stretches* both hands outwards *raising* them at the pace the volume level need to be expanded. The reverse action is true for the *decrescendo*. To play *accented/stressed* notes, a heavy hand gesture or stabbing action is used. Conductors may use characteristic, individual methods and gestures to convey their specific intentions. The music that listeners ultimately hear and enjoy, are the results of the interactions between the conductor and the corresponding responses by the musicians in the orchestra. This is quietly accomplished by the '*Sign language of conducting*' using manual gestures and body language!

In this ergonomically, and acoustically designed hall, several commonly found problems are vastly mitigated by an intuitive seating arrangement and height adjustable sections on the stage. The sections of the stage can be controlled by heavy-duty hydraulic lifts. The flexible capability allows for height, lateral, and slant adjustments of the sections.

In this unconventional seating pattern, everything is transposed by 12.5^0 horizontal *tilt* off-center and moving the conductor podium to the diagonal corner, where the '*speaking podium*' is normally situated. The feasibility and advantages of this unconventional, floating stage module design were elaborately studied by 3D virtual modeling and investigations. The individual platforms can be moved, horizontally, vertically, and angularly with flexible degrees of movement. Furthermore, they are constructed with telescopic cum retractable sections. Therefore, it may be adjusted almost seamlessly as demanded by each performance item.

The piano is moved further to the side and tilted in such a way that the '*cabinet lid*' is just only minimally invasive. It doesn't hinder view of the hind rows. If at all its sound is altered by a tiny bit, it may be electronically compensated by a skillful audio *soundboard engineer*. According to many, an open lid allows for clearer, louder, resonant tone to emanate from the exposed strings of the piano. They are also fitted with surface-mounted miniature cameras (as in mobile phones). All the playing parts of the piano, the keys, strings, hammers, and the damper which mute the vibrating strings, as well as the fingering actions of the piano player can now be clearly captured by the cameras. The wider views are shown by the hidden / camouflaged miniature cameras suspended from the acoustic shell dome /umbrella over the stage-deck. Furthermore, there are surface-mounted camera image sensors (in this case, the video signals processed externally) at strategical locations on other places on the piano itself to show the keyboard actions, mechanical movements, fingering, and the foot-pedals. The pianist's emotional expressions can be captured and shown in various perspectives. Another interesting feature is a smaller acoustic dome that can be precisely placed above a solo performer such as pianist, violinist, vocalist, accordionist, or a special instrumentalist. The placement of the shells can be pre-programmed and operated from the overhead grid transport system. Usually, they are placed at pre-determined locations; activated or lowered when needed, and smoothly retracted soon after use. In usual practice, all the device parameters are pre-programmed according to the music flow dictated by the music score, and the conductor's inputs. They are activated by '*function keys*' or switching console by the stage technician from his light /sound/ video booths as warranted. In the pre-production meetings, the related specialists go over the music /performance details together; the technicians annotate the required tasks accordingly for the deployment at the specified point in time. During an event, usually, the technical director or producer calls out the appropriate functional commands.

This major layout alteration will actually improve '*visibility*' and camera angles; reduce the usual performance restrictions imposed on the performers. The conductor's actions and the baton signals can be better viewed and understood by everyone- the instrument players, and the savvy viewer-listeners. The changes when initially proposed, were met with severe skepticism and criticism from the traditionalists. The mock performances conducted elsewhere in conventional halls incorporating the perceived changes, proved its overall benefits and a qualified acceptance thereafter.

The important idea is that the options are there if one likes to use them; if not, they are not deployed at all for that event. In other words, they are provided purely on an '*a' la carte*' (according to the menu) or '*as needed basis*'. Initially, the new technology and the various systems were extensively tested in amateur performances. Later, they were evaluated by the professionals for its suitability or suggesting the required modifications.

Most legacy auditoriums use a multi functional so-called '*catwalk*' which is actually an overhead grid structure with a platform for technicians to perform routine tasks. The grid provides a structure to hang lights, microphones, stage Loudspeaker, etc. It spans over most of the stage area and leads to the 'Lighting desk', a specialized section of multi-level technical booths. Lights are used for several specific objectives: accent lighting to highlight person or something significant;

back-lighting to sculpt and separate performers from the background; beam lights producing intense near-parallel beam; the dominant Key light, Fill lights to obscure shadows created by key lights; color wheel for dance performances, etc. There are electrical systems, cabling, plug points, dimmers, and a host of electrical & lighting gadgets installed and organized in this area with connecting '*bridges*' and walkways between sections. Sometimes, even during performance, there are movements and work conducted on the catwalk causing noise, distraction, and malfunctions affecting shows.

This area is usually dimmed and painted black, yet open to view by the front-row audience. These fixtures clutter and adversely affects many aspects of performance: sound, visual images, aesthetics. The numerous lights generate considerable heat and cause discomfort to the actors and service personnel. In here, these clunks, clutter, and blemishes are eliminated or replaced by smaller and efficient advanced technological-devises, capable of remote operation from the '*Lighting control board*'. Generally, incremental improvements in several associated components of a system make a noticeable overall improvement. In this facility, it's substantial improvements in every one of them, thus making a quantum leap and revolutionary impact!

A modern orchestra may consist of 100 pieces of instruments; usually, nine woodwind instruments, 10 brass instruments, 12 percussion and 50 - 60 string instruments. Seen from the audience perspective, the orchestra is organized in a semicircle pattern in the order as the strings in the front, woodwinds in the middle, and brass and percussion at the back. The choir is normally placed all the way back behind the orchestra on stage. The solo voice singer is at the front. The higher pitched string instruments are seen on the left and the progressively lower pitched instruments moving to the right. The piano sits the furthest left of the violin group. All are facing the conductor so that each can receive cues and commands from the conductor as when to join in at what *measure* of the music score and when to exit; the pace, tempo, dynamics, and other fine attributes of the musical rendition.

There are certain biological and acoustic considerations dictating this particular organization. The biological function is predicated upon the fact how brain process the information it receives. It shows a '*cross function*' peculiarity, meaning the *left side sensory* data is processed by the *right cerebral cortex* (outer layer) and vice versa. The acoustic consideration is related to the loudness level of the instrument groups. The sounds of the percussion instruments are produced by the forceful striking actions and emanate as '*impulse forces*' (A large force acting for a short time). They are quite loud; so also are the blaring brass-wind instruments '*beaming*' their loud sounds, ejected from the bulging *bells*. Both these instrument groups are placed higher and in the back; percussions in the very back on the left while the chorus to the right side. Some of its sound power is absorbed by the bodies of instrumentalists sitting in front mitigating the harshness, thus bringing a semblance of balance with other voices in the orchestra. On the contrary, in Indian music, the percussion ensemble (Tabla, drums and group of instruments) is seated in the front row towards the left. The brass instruments are almost totally missing, except perhaps for a saxophone for a rare, brief solo, or as a short fill-in accompaniment. The accordion is also sparsely used; usually supplanted by an electronic keyboard producing *synthetic* (artificial) sound. In case an accordion does appear here and there, it merely plays the melody variations; the bass cords are not usually played, yielding the role to the percussions. The Indian *thala* sequence and the accordion chord setup are not smoothly compatible. *A music cannot have 2 separate rhythm patterns at a time, although melody can, as in 'counterpoint'!*

- A word on the Piano

Gracing the stage, is always found a grand Piano, ever ready for its use. It's a Steinway, a shining black masterpiece. Piano has the broadest range of any *stage-instrument*. Therefore, music for all other instruments can be composed on it. It can be played for solo, or virtuoso performance; symphony orchestra, concert, or for accompaniment. A piano has more than 11,500 individual parts. Associated with each key, there are about 130 parts and many essential common items. Originally an Italian invention, in the very earliest period of the 18th century, the Pianos are made of the finest materials with meticulous craftsmanship for excellent sound production and appearances. They are made from maple, poplar, or birch for the Case/cabinet, shaped to conform to the layout of different lengths of its strings which have a '*nonlinear*' proportionality [the longest string on the extreme left; then becoming progressively shorter to the right side of the piano player]. A modern grand piano has 88 keys the lowest sounding note A_0 resonates at <u>*27.5*</u> Hz (just above the lowest human hearing limit of 20 Hz; most people don't hear it; But it's there for the very few gifted!) and the highest and the last note is ***C_8*** at <u>*4186*</u> Hz. Altogether, there are 8 *A, B &C* keys; only 7 D, E, F, G keys. It has a total of *52* white keys and *36* black keys. The 49th key ***A_4***, or simply the '*middle A*', is tuned to *440* Hz,(2^4 *27.5 or simply 16*27.5) the reference note for the tuning of all other orchestral instruments!

Slow-grown, Alaskan ***Sitka Spruce*** wood, that is close grained, fine, even textured with consistently straight grain pattern is selected for the most important and highly sensitive *soundboard,* sometimes reverentially called the *'tone-wood'*. (connoisseurs of music and luthiers call sound, a "*tone*"!). It has good strength and most importantly, the elastic quality to vibrate appreciably and adequately! The paper thin (comparable to the violin shell wood material. Spruce is a softwood with high hardness and is used for the top part of the violin!) resonant sound board acts as the '*hi-fidelity*' amplifier of the feeble sound produced by the vibrations caused by the strings. A typical grand piano has a total of about 236 strings. The lowest 10 bass notes have only 1, but heavy string for each note. The next 18 have 2 strings / note, and the other 60 have 3 strings /note. The fact explains why 88 keys have much more strings than the number of keys! More strings for the higher notes should naturally provide louder and richer tones to the higher treble notes, besides giving a balanced sound level relative to the bass keys. The sound vibrations travel along the grains and to the center. Two laminated *bridges* of hard wood *Maple* fitted with bridge pins, one a short bridge to hold the Bass strings (due to the fact, they have only fewer number of strings); the other, long for the tenor, and treble strings, are glued to the top of the soundboard. They transfer the mechanical / kinetic / vibrational energy of the strings to the sound board. This is analogous (functionally similar) to the vertically mounted, thin and slotted, bridge linking the 4 violin strings transmitting vibrations to its hollow body chamber of air column.

Piano strings are specially made from tempered high-carbon steel called spring steel, possessing high tensile strength of certain critical *gauge*. They are cut into specific lengths, corresponding to its wavelength (frequency is inversely proportional to wavelength by a constant factor '**c**', the speed of the soundwave in the given medium which in air at room temperature is *343 m/ sec*. The formula in Physics $\mathbf{F} = \lambda/\mathbf{c}$,) of the note. The diameter and the length of the strings increase as the pitch or frequency gets **lower**. So, the bass strings use a single but much heavier string which vibrate in its lower natural vibrations. In the extreme bass, the strings may have a double winding of copper. The bass strings run across over top of the tenor and treble strings. The treble pitches which are actually high frequency vibrations of feeble low intensity have three unison strings (two or more strings that are tuned to the **same** pitch), the tenor range uses two unison strings of medium thickness. They are tightly strung and run linearly over the soundboard. Then, there is the pin block and the cast iron plate of '*Bell quality*' providing structural, alignment and tensile support for the 236 or so taut strings. The whopping string tension range from 70-90 Kg/cm^2. The precise tension is essential for consistently producing the precise number of vibrations of the notes.

Key / hammer shanks are also a critical component in the transfer of energy from the fingertips of a player to the strings on a piano. Piano keys are made of softwood (conifer) varieties. The keys, a dual lever arm which pivot at the *balance rail*; are padded with *felt* to prevent rattling noise. They are covered with black or white plastic. The striking Hammers belonging to each key, have a core of wood molding with under-felt and top felt both molded / wrapped to the core with *resin*. The thinnest felt is used for the treble hammers, while thicker felt for the bass strings that are much heavier needing larger energy transfer to cause the required vibrations of the strings.

Tuning the sound of the piano is accomplished by a specially trained tuner called the '*voicer*' who mostly does his job by critical listening. Many, now use modern, integrated sound / frequency / level meters. My onetime colleague and TD *(Technical Director, video switching)* Maria's husband was a voicer. He declines all electronic aids in tuning, claiming "*there is much more in the sound than the device can tell me!*". I, who often miss syllables in conversations had intently watched in perfectly held silence when a tuner was doing the tuning in meticulous listening rituals. In another situation, I did the same when a person was performing '*reed tuning*' for my accordions. Unlike the Piano tuning, reed tuning for accordion has to be performed within the '*tone holes*' themselves, and in many tedious iterations. It was a curious sight for me to see how repeatedly and critically they listen in both cases. In piano tuning, the tension of the strings may be tightened and/ or the felt hammers modified by *re-texturing* the hammerheads, altering the sound. Many technicians remove material (mass) from the *hammers* to achieve more sustain and to impart more of a mellow '*singing*' quality to their pianos.

When playing a piano key; the hammer strikes the strings; causing vibrations; it's vibrating / kinetic energy is spread through the bridge and onto the soundboard; the soundboard vibrates almost invisibly, agitating and exciting the surrounding air column which in turn *compresses and expands* cyclically under varying pressure producing sound. How skillful and ingenious! It is the sound board that projects all the sound waves we hear and enjoy so much! A grand piano has commonly 88 keys. An Austrian made *Bösendorfer* grand Piano is said to have 92 keys. I find the 88th key too sharp and annoying. Thankfully, in most music a note like that is employed as a transitional one of short duration or in '*tremolo*' (a tremulous effect produced by rapid repetition of a single tone or rapid alternation of two tones).

The stage floor is tiled with selected wooden panels chosen for their acoustic qualities and physical strength. They are delicately laid, on heavy metal beams for structural support, and carefully avoiding any squeaking noise. The beams themselves are much less sensitive to sound waves. There are arrays & bundles of color coded, and labeled cables laid underneath the stage floor in separate soundproof plastic conduits pre-installed across its length and breadth. They include fiber optic, video, audio, communication, MIDI, and data cables with a myriad of connectors and pin configurations. They surface to the floor only at appropriate *plug outlets* at desired locations for facile and tidy connectivity for Mic, Instruments, Electric power and many other requirements. The '*blue-tooth*' (a wireless technology standard for very short-range data communication) and '*Wifi*' (a wireless digital/Internet networking protocol for Internet connections of digital devices) are truly blessings which reduce if not eliminate the clutter, improving the sight and safety on the stage saving many trips and falls.

It is noteworthy to mention that there are no music sheets used in the musical performances. The instrumentalists, vocalists, and the chorus singers are provided with easily legible *Ultra HD resolution OLED* display tablets on telescopic mono-pods (stands) with swivel/tilt heads. The music sheets are shown on the display panels. The page turns occur with a touch on the screen, or by a foot switch. It is not necessary that every instrumentalist has to be provided with his / her own music sheets. They simply share a panoramic screen with wide viewing angle. The soloists get their own individual tablet display. The traditional music stands are '*tripods*' with three telescopic, and collapsible legs. They are visually intrusive, cumbersome, and take up wide floor space. These innovative arrangements reduce the '*on stage*' muddle of fittings and gadgets; at the same time, improve a great deal of visibility of performers, their instruments, and playing styles! Every voice source whether its vocal or instrument is systematically arranged so that the surround sound system indicates the location and the depth of the voice source on the stage. The violin plays at the left and front, Choir at the central-right at the very back, and so on.

The backstage provides plenty of real estate. It accommodates, Instrument storage & rehearsal area; green / makeup room; costume racks, segregated rest lounges; and a curtain /props/ riser platform/ gadget room. All designed and located after considering many factors: acoustic, access, ease, and convenience of rapid deployment. There is wide loading dock and ramp access to the backstage. Hydrogen powered lifts, hoists, and machineries are used in safe and speedy material handling tasks.

Eminently resourceful Audio console where sound is processed

Audio booth is situated by the right side of the stage, partitioned with large lookout '*one-way glass window*'. It is the most suited location for the operation. The wood paneling scheme requires that shorter panels be accommodated on the side walls. The booth houses the main audio mixer consoles and accessory devices. Many other audio equipment such as sound amplifiers, equalizers, signal monitors, PA system, patch panels, etc. are installed in the booth. The audio mixer console is a sprawling control board with 96 audio input sources, with many '*sound enhancing*' controls and mixing of sources with lever faders. It takes usually 3 specially trained sound engineers to operate. The PA system is equipped with its own separate switching & control console operated by PA system engineer.

The *VideoProduction Booth*, is situated all the way back up on the top tier of the balcony. It occupies a large area, projecting outwards giving a vantage view of the stage and the entire seating area of the hall, tilted enough to give a good long view of the audience. It is the main hub center where all video signal sources terminate as '*inputs*' and emanate as '*outputs*' for broadcasting and all the display panels in the surrounding complex. A variety of signal processing functions are performed here. It houses a pair of grand "*GrassValley*" video switching & control consoles and a host of auxiliary equipment.

The multi-functional video- production system where video signals are processed

They are highly versatile and powerful electronic devices; manipulate pictures; create visual effects of many kinds. They can artificially add fog, mist, smoke, background scenes, reflection; crop, scale, stack images; correction of various video parameters, etc. limited only by the imagination, time, and capabilities of the engineers operating the console or by stringent production requirements. The engineers manage the complex operations by partitioning it by functions, pre-programming operations, and by automation of recurring functions. All cameras (at least 120) are connected to this booth. A team of 3 qualified T/D perform select, mix, match, and packaging operations in different ways, before putting the signal '*on air*'. Everything is based on Ultra High Definition TV formats, norms, and standards. The chosen standard picture aspect ratio is 16:9 (international standard format of HDTV). There are '*body cams*' and cameras placed on instruments, providing their *perspectives* of the instrument and playing style. For example, there are inconspicuous surface-mounted cameras strategically placed on the piano; looking at the keyboard actions, pianist, strings & hammer motions. In this hall, it's paired with a programmable system controller. It can send *composite* signals to the video control booth as programmed; composing or packaging images in various collages, patterns, sequences, and 3D perspectives; sometimes showing simultaneous actions in 2, 3, 4 way splits. Consequently, the control room operators can perform their operations faster and efficient.

Concert is no longer an audio only performance, although sound is the most enjoyable part. They are many things to witness and experience in a live performance. People flock to the concert wanting to enjoy the full package of actions, playing styles, conducting, and feel the vibes, aura, atmosphere, acoustic qualities, and the philharmonic performance. Many considerations are given to provide the audience and viewers a true '*immersive*' audiovisual experience! Background images, picture collages, stock footages, animation, and 3D displays are intermixed and displayed in real-time during the performance.

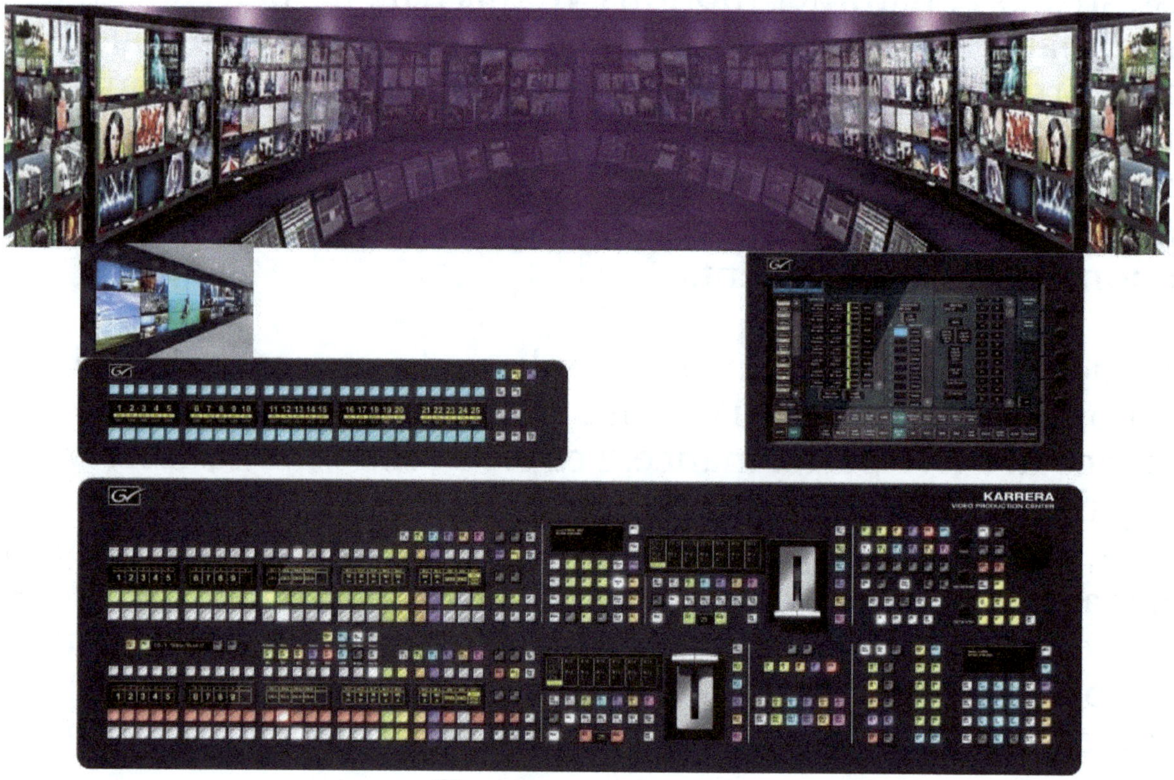

The video wall & the control consoles

The video wall is an array of large curved UHD (Ultra HD) resolution OLED (Organic LED) display screens fitted on a frame of 8m X 4.5m in the standard HDTV format of 16:9 *aspect ratio* (length x width). It serves as a background for the stage. Usually, selected still & motion video images of the events are scaled, collaged, juxtaposed, and displayed on it. It comes quite handy for showing various visual effects of close up, split images of several simultaneously occurring scenes.

There is a similar but spherical Roof display panel, hung over the stage floor, and tilted to towards the viewers having the same facility and features as the video wall. It can be retracted back into obscurity if not needed. It is mainly intended for movie productions. It could realistically show, in a very convenient way, the skies, sun, stars, moon, Eclipses, clouds; aircraft, birds, projectiles & missiles, Aliens from other planets; Flying Saucers and meteors darting across the skies, and for the virtual 3,D effects in science-fiction movies.

The video wall has its own integrated audio system using different digital streams for a specific screen and Title generator. The Title generator is used for titling purposes to name the presenter and guests with their official designations, status etc. and to generate crawls and 'end of the show rolls' to show names, titles, performance rolls of actors, stage crew, and contributors in their specific fields. It's especially useful for showing multiple and/ or simultaneous views either from different locations as in the election results-coverage from different states, or from different camera shots as in sports events. The viewers get a total view of several actions that is taking place and choose to watch and hear the most interesting one. Each screen is capable of showing captions and other details of what it is showing. Theoretically, the viewers can miss nothing if only we could use our pair of eyes properly. This is a great tool of any kind of presentation as it may be directly controlled by the presenter himself/ herself from a single system console, or by a suitably pre-programmed 'interactive display menu'. Its applications are many as in education, healthcare, entertainments as often seen here at the Art center performances. Its impact is huge and effects stunning! The video wall is mostly used to provide a multi-image background screen or direct view at major 'games-watch parties' at stadium's affiliate venues and city squares for large public viewing.

In practice, any images can be switched and shown on them. The Video wall is operated from the Auxiliary building by trained engineers and Special effect '*tech-artists*' using full-featured video switching consoles. They are prompted by a T/D present at the auditorium with a large view of it. The Indian '*Movie studio association & Producers Guild*' invested $ 45 Million in the whole project to incorporate many film production facilities & amenities to the complex. The old-fashioned and the traditional-thinking Board of directors wanted all their movie scene props and other gadgets transported to the Amphitheater complex. It resulted in some disputes, and acrimony threatening legal actions. They were oblivious of the power and capabilities of the digital domain.

By sheer chance, the technologists came to know about it and resolved the problem instantly with a simple digital solution. The technocrats (technical experts / managers) went to each of the Movie studios located all over the Film cities in India and took the inventory of the props. Each of the prop was tightly stretched on a large mesh frame without any wrinkle and laid them flat on a large parking lot. They lowered a crane with multiple cameras mounted from it, and over the props. Then, took hi-res pictures and digitized them on to tiny micro 3D NAND memory cards. Each SD card holds up to 120 Hi resolution, uncompressed images. All props used from the ancient times to the recent day were digitized in the same manner. They are now stored in the servers at the complex, instantly available for use over the encrypted, and secure networks with top '*Network Security Certifications*'. They bear digital signature printed on them which may be totally erased by encrypted technology and method by certifications from the hierarchy of managers. Most interestingly, the complex have the capability to alter them according to any special needs. For example, the same scene could realistically appear in different environmental conditions of daylight, evening light, sunset glow, romantic moon light using '*image manipulation program & digital processing*' applications. The sea may become turbulent with storm waves; the trees may swing violently in thunderstorms; the nonvocal may erupt; planes may land at the edges without being burned and so on. The effects are cataloged, labeled, and made readily accessible from pre-programmed soft touch template icons. In modern times, it isn't Film production any more; it's '*end-to-end*' Video-movie production! There are video projectors now available with resolution and brightness specifications superior to those of film projection. Everyone is pleased with the results.

Moreover, by the technique called '*chroma keying*' in which any uniform color background the camera is seeing, could be removed and replaced by any other static, motion video, or animated background. They may come from files of slides, or 'stock-footages', or created animation for the specific purpose. There are other creative methods and techniques that can be seamlessly integrated. Anything that cannot be accomplished by human capabilities can be achieved by the proper use of hi-resolution and motion compensated animation. Done properly with '*motion compensation*' methods, they provide a very realistic look and feel of any required scene! The same familiar technique of '*cut / copy & paste*' from the word processor programs can be applied here too. The "*stunt performers*" do all the adventures. The "*Visual effects*" teams make magic with their skills and the technology they are provided with, to bring about "*virtual realization*" in movies. They can make movie stars look realistically whatever they want; create situations the directors ask for; show the hero's hands bleeding, face disfigured, shot at, jumping off from airplanes; climbing on speeding trains; rescuing trapped victims from nuclear power plants, or out in the seas when the storm is raging threatening to overthrow the sailors, just using the replica of the vessel floating in a swimming pool or large open tank with a green background. The virtual reality can make anything look a real '*reality*'! Collectively, they are called "*special effects*".

The stunt-men and specialists work in the oblivion in the backstage or background. The heroes or the protagonists win all the rewards- money, acclaim, accolades, awards, and fame at the expense of the '*substitutes*' who really perform the hazardous tasks, and yet live poor. The stars always steal the show and everything else that comes along with it. They '*shine in borrowed plumes*'! This is what is happening in the fake '*make-believe*' world of Movie, Television & magic.

Quite ingeniously, the large dome is equipped with layers of light sensitive, remote controlled transparent / translucent / opaque, and sliding convex shutters, that can unfold like flower bloom in spring, opening the petals, only much faster at the dome. A zoom camera pans and scans the skies; display live pictures on a large background screen; the opening let the ambient lights in and casts on the characters on the stage. This feature helps to select the '*ambient*' light and the '*sky view*' which is also very useful in the movie productions. In addition to this complicated process, there are lights and lighting technics that can be utilized effectively in similar circumstances. Familiarly, daylight are bright white lights possessing wider light spectrum. The LED daylight bulbs mimic natural sunlight. A higher '*color temperature*' in the range of 5000 – 6000^0 Kelvin, makes '*daylight*' that casts a white hue with blue tint. Then there are '*warm white*' and '*cool daylight*' suitable for different times of the day, distinguished by a mix of hue, and tint. The intensity of light is measured in the unit of '*lumen*' that is measured on a light meter for proper setup. The amount of light falls on an object can be adjusted by switching on and off of different light sources. There are fixed and portable light fixtures, a combination of which can provide, highlights, accents, color, shadows, etc.

Once invested in digital technology and applications, movies, serials,(Published or produced in installments, as a novel or television drama.) sequels (Something that follows as a continuation, especially a literary, dramatic, or cinematic work whose narrative continues that of a preceding work.) can be made without massive budgets; needs only minimal location shooting. Even that can be accomplished by taking pictures of the location and later manipulating and interpolating as needed on the movie set, or in post-production edit suites. Low budget producers can also participate and benefit by renting the production facilities along with the technical crew. They may order what they want *à la carte*!

Initially, the amphitheater can be used as a multipurpose, multimedia, and concert venue, as well as Television & Movie Studios. They can learn and apply the findings and knowledge in the '*design & build*' of another solely dedicated facility for Television & Movie making. Expert editors and post-production staff will '*serve & deliver*' what are ordered, and desired! Altogether, the amphitheater provides an '*immersive & augmented*' listening pleasure; exceptional visual display, and an exciting experience to all participants- the audience & performers alike!

The pompous dome of the concert hall situated on a typical suburban and bucolic complex

The complex sits on 57 acres of landscaped, mowed, and manicured meadows, with a beautiful garden of flowers, artificially made waterfalls, aquarium, and children's park; gazebos, pavilions, patio porches with wicker furniture. The lawn is weekly mowed and trimmed with subtly contrasting geometric patterns. The patterns change every month. One of them, is a familiar pattern of concentric green circles as seen in the '*bore well irrigated*' agricultural farms around the '*Wadis & Oasis*' in the Sheikh's own native country and easily noticed by air-travelers. They are further beautified with custom landscape lighting. The winding pathways in the park alone is over 3.0 km long. It is compared with the "*Brindavan Gardens*" in Mysore.

At the Eastern edge, away from the Amphitheater, is a '*helipad*'; its perimeter is illuminated by a chain of tiny blue, green & white LED lights. A '*windsock*' tells the pilots which way the wind is blowing and at what speed, shown in figures on a cube shaped sign-board at the tip of its mast. It orients itself against the wind when the wind speed > 3 knots. At a wind speed > 15 knots, the windsock is fully extended and points to the opposite way the wind is blowing (windsock pointing due north indicates a southerly wind). It's illuminated at night triggered by a light sensor. It assists the pilots with a quick visual check of landing pad and the conditions for safe landing or avoid it when strong tailwind blows.

There is a spinning yellow beacon in the middle, mounted flush with the landing pad. When the Helicopter arrives, the turbulence and the mushroom draft is detected by sonar sensors at an area of 22 m in diameter and a height of 25 m, further assisting the pilot with a safe and pinpoint landing. After the helicopter has landed, bright lights light up to help the passengers disembark safely. The service is suspended during performances, except in medical, fire, or any other emergency situations. The service is mainly used for transporting the top government officials, celebrities, diplomatic officers, or special invitees. It takes merely 7 minutes to the Kochi International Airport, which services a passenger traffic of about 10 Million with operations of about 40 daily flights. It prides itself as the '*world's first fully solar-powered airport*'. Besides, it is one of the first public / private, national / international business ventures of its kind. Unlike many airports in the world around, Kochi airport sports a '*ranch style*' rustic terminal building reflecting indigenous architecture, shape, local settings, and charm. It's a sprawling complex of serenity and calm, unlike the usual vertically integrated, rising mega structure symbolic of a crowded, host city with hectic lifestyle and congested streets. The homebound passengers enjoy a homely welcome when they land.

The Malayalee diaspora had long been suffering from the notorious and dreadful conditions at the '*port of entry*' at Bombay, where corrupt airport /airline officers, custom officials, smugglers, hotel, tourist brokers & taxi drivers, as well as money exchange scammers & scoundrels have bullied, harassed, duped, and stole goods & valuables from the weary travelers. At one time, an Indian Airlines ticketing manager demanded US Dollars only, to buy tickets for a flight to Kochi. I asked her why she cannot accept Indian rupees, in India, from an Indian citizen for a domestic flight; and that she is in serious violation of the country's currency laws. These nasty tactics will only cause the value of Rupees to deteriorate! This is just one of myriads of horrible experiences. Passengers could not experience anything worse than those they were subjected to at Bombay in those miserable days. They uttered curses, swore profusely at every obstacle, and inconvenience unnecessarily thrown in their way. They would rather want to go back to the same place they came from on the very next return flight, but for the undiminished love for their immediate family, the homeland, the food, the salubrious settings of a home, the affection, and company of the loved ones.

On the contrary, the home airport at Kochi International now offers a much craved haven and a loving welcome by the immediate members of the family who have traveled a long way from their respective villages to here to welcome the loved ones first hand, and warmly escort them home. Mothers & fathers, wives, husbands, children, relatives, and friends are among the entourage. All have arrived well in advance, and been anxiously waiting for the timely and safe arrival of your flight! It's always a rapturous welcome of relief and affection soaked in tears of joy on every face.

The sprawling and fabulous Art center complex

There is an artistically and artificially made pond on the far side of the park complex, shaped like the map of India. The state of Kerala is shown as a floating island in the Arabian Sea, drawn to the bigger landmass of India, poignantly reminding of its mythical origin. In the beautiful reflecting body of water, float the replicas of indigenously used vessels & boats of various kind; cargo sailboats, explorer & merchant ships or ships built for spice trade; navy ships and Aircraft carrier '*Vikranth*'; the colonial war ships; the race boats- Chundan Vallams, and other boats, with models of palatial bungalows along its banks. The model rekindles authentic historic reminders. The boats in the ponds are slowly plying between various destinations. They navigate '*autonomously*', controlled by electronic modules with miniature, programmable, '*self-driving*' microprocessors in them (advanced chips of similar kind may soon find a place in autonomous driving vehicles). They go around the tagged posts, marked by colored flags of each of the states. It is quite amusing to watch how they maneuver and negotiate their routes; they are equipped with '*proximity sensors and collision avoidance systems*' in them; they speed up or slow down, as needed to avoid accidents. At 6 pm on the hour, all the floating vessels assemble as an armada at the foot of the National Flag for a 5-minute ceremony. The national anthem is played in its original fanfare. At the other end of the pond is an awesome replica of the '*rocket launching tower*' standing ready to fire "*Chandrayaan*" space exploration rocket. After the playing of the anthem, the 'countdown' begins; an oozing sound is heard while some thick white cloud of fumes spews out of the base of the rocket engine. It is immediately followed by a sonic sensation that the rocket is being launched by the sound effects created by a series of loudspeakers mounted at proportionately varying distances on the tall '*launching tower*' progressively attenuating (decaying, losing intensity) sound levels. It imparts an authentic sensation that the rocket has been launched! The blinking arrows on the tower mimic its upward path. Everything is synchronized, choreographed and automated. It's a favorite spot for children, families, and the visitors. The school children join in singing the national anthem in solemn awe, standing in '*attention*' just like in the regular morning '*Assembly & salute*' ceremony conducted in their schools.

They shall never be "*divided, and ruled*"; nor "re-*colonized*" again! The ceremony ends with a solemn pledge.

There are a number of plaques on pedestals with basic civics lessons inscribed on them; some golden principles of ethics and morals are engraved in golden letters. Parents, strenuously make the children read them; paraphrase (restate using different words, especially for clarity) the gist of each statement afterwards inculcating into their consciousness. The youth must grow up as civic minded, responsible, and dutiful citizens, respecting all others, regardless of caste, creed, wealth, stature, or social status. It's necessary to practice them for the sake of unity and peace in a secular country. All are children of God and created in equal love!

At the setting of the dusk, flashing tiny lights comes on the perimeter of the pond and cast various patterns of colored light beams. The reflecting lights modulated by small waves scatters the light; offers different perspectives at different angles. Some days it may look like a silver cascade of falling water; another time, a molten lava flowing down. It creates various images of climate changes; depletion of arctic ice, falling avalanches, floating and drifting icebergs in the water, penguins parading in unique formations; stranded polar bears howling desperately from hunger; trees devoid of leaves: sand dunes spreading across, invading fertile agricultural lands, and crying families of desolate farmers. They are dramatized and shown by special overhead video projection coming from different angles and elevations, programmed and presented on the surface of the reflecting still waters. Each scene is accompanied and accentuated by proper sound effects. It instills environmental concerns and consciousness among the public! It helps to keep the surroundings and public places clean, and tidy; streams, well, and water safe for the people to use and protect marine life. Qualified volunteers explain horticulture, crop cultivation; maintenance of gardens, orchards, and farms.

There are two other buildings situated in the property. One is a Post Production facility. That is where the Event recording, Editing, Graphics, Computer art & Animation, as well as other Multimedia processing are carried out. This is quite a departure from the conventional design where everything was concentrated. The light-speed networks make the '*distributed*' operation reliable and possible. The auxiliary centers serve discrete functions. There are separate computers, Digital, and Video streaming server clusters & racks. In addition, there is a fiber optic "*Trans-code & Link Assembly*" Control room, and a Telecommunications center. The complex, sometimes termed *campus*, is served by fiber optic network with an array of single mode, multi-mode, multi-wavelengths fiber cables driven by Ruby and other '*Laser*' beams for carrying audio, video, data, telephone, Intercom, signaling, and other communication signals. During the performance of events, signals travel back and forth between the buildings in real time at the so-called light speed, 100 Million times faster than Usain Bolt's sprint, 5 Million times faster than the Agni missile, as many times faster than the arrow shot by the Swiss legendary hero William Tell, to show some mundane analogies, and comparisons. In layman's term, it happens faster than a blink of time! All the modern cable networks, whether terrestrial or marine, have adopted this technology for Internet data, Television & video streaming, and other image, and communication signals in their chosen digital (in bits & words) formats.

Well, the critical idea is that all required signals are available instantly when needed at the desired locations for the devices that depend on the media contents, external commands, and/or for switching. International News agencies and Broadcast stations access news / sport video footages in real time from '*cloud servers*' located overseas, via undersea fiber optic, or satellite networks. It's vital for international news organizations who broadcast to various time zones to package and update news bulletins true to the current status. The same trusted and time-tested technology is adopted here too.

The most interesting fact is that the applications of the power of light beam, use of prism (a crystallized structure with a property of focusing the light beam and harnessing the power of light rays) have been well-known for ages. In fact, it has been in use from the time of ancient civilizations. In those days, the primary use was to make fire or melt hard metals in the process of making tools. Here, it is pertinent to mention that in one of the episodes of '*MacGyver*' TV series, named the '*Legend of the Holy Rose*' the very same principle of concentrating light beam from the sun is demonstrated in finally solving a legendary puzzle!

A third building in the complex, is the Administrative building, where the executive suites and offices of various departments, Engineering & Design/ Maintenance are located. The Operations dept. is entrusted with the smooth running of the events- Logistics, Training and Public relations. All the positions and personnel were hired almost 15 months ago. Since then, they were at work. Each one of them was given exhaustive apprenticeship in their assigned subject area.

Each technical and support service personnel is appointed for the specific art form and performance to be enacted based on their comprehensive knowledge and demonstrated skill level in the area. They were given rigorous training in theoretical, technical, and operational matters by teams of world renowned experts and technical trainers from each of the original equipment manufacturers. At the end of the course, exams were conducted to prove their proficiency, knowledge, and judgment in a fast-paced '*live*' event environment, before they are inducted into the Operations cadre. Following a successful completion of the training period, the technical staff conducted numerous rehearsals to prove operational readiness; remedied errors, or mistakes that befell in the process. The readiness report was then presented to an external body of carefully constituted, competent experts, and critics for the "*proof of performance*".

Mainly, there are two art forms:-Fine arts & Performing arts. Fine arts are deemed as the highest standard of artistic expressions. The main function of Fine arts is to create aesthetically pleasing works. They are mainly visual and static art works; while the other artforms are dynamic performances in front of audiences. So, fine arts should include any work involving drawing, painting, sculpture, ceramics; architecture, literature, etc. In addition, there is also a recent classification called studio art.

The Performing arts include physical activities such as drama, theater, music, and dancing. Sometimes, the latter two are considered '*high arts*' and therefore "*fine arts*" too. But, all physical activities such as sports & games, martial arts, etc. are not considered arts. They have other befitting classifications.

In summary, traditionally and historically, the five major fine arts are the painting, sculpture, architecture, music, and poetry. Drama and dancing are considered minor arts. In the contemporary culture and discourses, the visual arts and performing art forms, such as painting, sculpture, collage, assemblage, calligraphy, music, dance, theater, architecture, film, video, photography, conceptual art, and printmaking are all included in the same classification.

'S*angeetham*' & '*Natyam*' (Music & dance) are considered divine in many oriental cultures, as practiced in liturgy, '*pooja*', chants, hymns, '*mantras*' (word, a syllable, a phrase or a short prayer that is repeated over and over again in a chant), and prayers in expressing devotion, veneration, and praise of God. '*Deepam*' (oil lamps) candles, incense burner, Holy water sprinkler, statues, icons, and relics symbolizing, the five primary elements- Fire, Air, Earth, Space, Water, and Wisdom, provide suitable reminders of godly presence. There are '*Natyam*' steps and dance movements specifically intended for worship. It is clearly visible in "*Vaishnavism*", an offshoot of Hinduism, as well as in the recently popular '*charismatic*' prayers among Christian congregations. It was more commonly enjoyed outside in connection with religious festivities; as adjunct activities for the public to drum up interest, and provide a platform for teaching the religious dogmas and doctrines with the aid of the entertaining art forms. Now they are becoming a part of religious services. The Christian rites and rituals are gradually drifting away from the total domination and extreme stronghold of Latin influence, and culture.

Studio art refers to a commercial / business oriented graphic, animation & computer arts; AR & VR (augmented & virtual reality): Television advertising materials such as Film / Video / Photography, cartoons and poster art; Billboard signs, Big screen '*signage*' (The design or use of signs and symbols); signs, posters, fliers, comics, graphic novels, street graffiti, etc.

Ballet, is a performing art, a theatrical dance combined with other artistic elements such as orchestral music, costume, and stage scenery. This dance form shows off the amazing and elegant body movements, flexibility of limbs, aerobic movements. The poses have their own genre of terminologies; distinctive feet positions numbered 1-5; body rotations at the hip; bend, stretch, hold; pivoting only on the front toes, etc. Pyotr Ilyich Tchaikovsky, is widely recognized as an avid proponent of this art-form for his ballet music compositions such as '*The Sleeping Beauty*', '*The Nutcracker*', and '*Swan Lake*'. Ballet is sometimes introduced as an adjunct (a dependent but a subordinate addition) in an Opera.

The covenant (a binding agreement) formulated by the cultural ministry of India stipulated in many rules that:

- No ballet performance will ever be performed in the hall.
  [It is considered as culturally too offensively explicit, vulgar, voyeuristic, and revealing!]
- 3 out of 5 performance shall be pertaining to Indian heritage, origin, and creations
- No alcohol will be served
- No smoking is permitted
- No political, religious, or radical themes will be staged
- All usual censorship rules about nudity, vulgarity, profanity, indecency, blasphemy, and related norms shall apply!

The concert hall is an '*all- consuming*' passion about the sound occurring within the human hearing range in its sublime beauty and greatest excellence. Essentially, a sound is produced when varying number of vibrations are made in the air causing continuous changes in air pressure for any duration of time. In music, we refer the amount of number of variations as "*pitch*" and its differences in '*notes*'. Every music note has its own specific pitch. For physicists and audio engineers, it's "*frequency,*" measured in the unit of Hz and the differences as '*frequency variations*'. The vibrations permeate all around: front, back; up and down in the 3D space in varying degrees of intensities depending on the distance from the source.

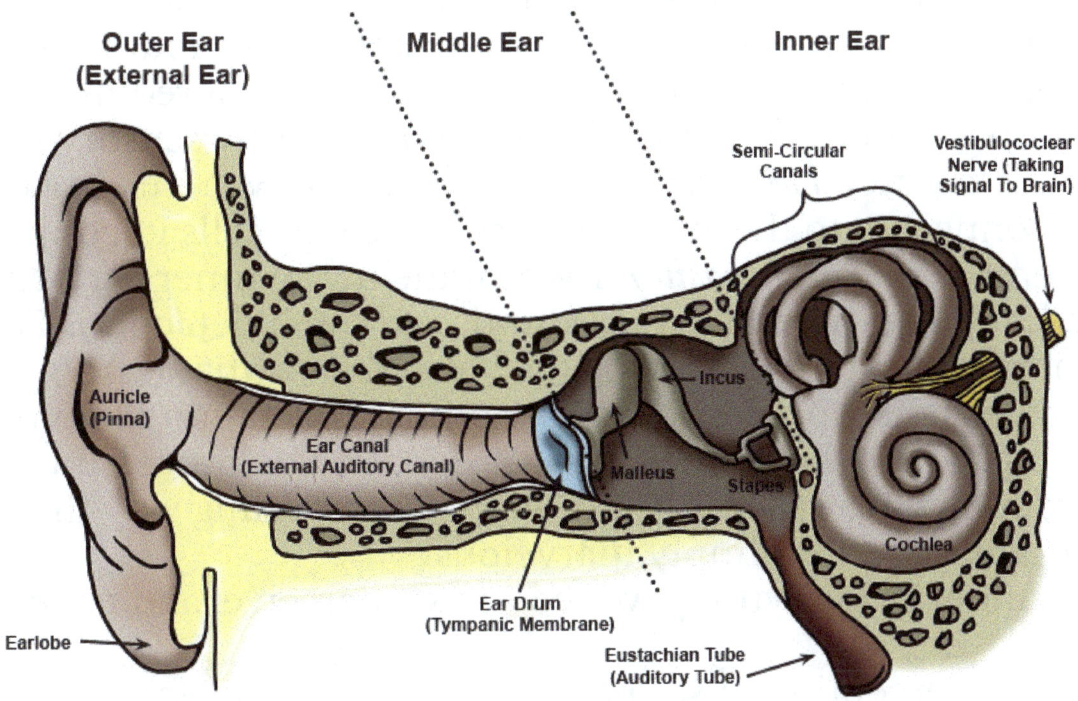

Anatomy of human ear (transverse view)

Sound is received by the auditory sensory organ, ear. The '*outer ear*' through the '*auricle*' (the outer projecting portion of the ear, also called 'pinna'); leads it through the '*ear canal*'; on to the '*middle ear*' which is surrounded and protected by 3 bones. The ear canal is like a tube, open at one end. So,

it has certain natural resonant qualities in the range of 1 k- 4 kHz. Possibly, that's why we, humans like the sounds in this range so well! The eardrum is located at the end of the ear canal at the entrance to the inner ear. The ear drum acts as a mechanical device as well as a '*high pass filter'* allowing the higher frequencies (greater than 600 Hz) to pass through easily, and at the same time, attenuating or discriminating lower frequencies. The ear drum connects to '*cochlea*' (A spiral-shaped *cavity* resembling a snail shell, importantly, containing '***nerve endings***' essential for hearing) in the '*inner ear*' via other mechanisms. Here, the sound vibrations are converted into '*neural information*', (the only language brains seem to understand, as the computer understands only *bits!*). Nothing shall take place in the body without the vital participation of the brain! Further, the cochlea performs Hi-to-Lo filtering similar to "*Fourier Transform*" (a mathematical operation that can break down *a complex signal into its constituent frequencies and intensity levels. It's amazing that organs know mathematics even if we don't!*). It may be doing so to assist the brains in processing the information faster and in real time! That helps '*synching*' the visual information of mouth movements of the speaker with the hearing by listeners. The gathered neural information is sent to the brains by the action of tiny sensory cells called inner & outer '*hair cells*' (named so, perhaps, because they are slack, tiny, and shaped like strands of hair) connected to the '*cochlear nerve*'. The hair cells in conjunction with a membrane called '*basilar*' membrane in the inner ear perform advanced, sophisticated filtering, sound, and pitch detection. The central nervous system learns what is desired and what is not; discriminates noise; process the vital audio information from the content to produce the '*processed*' aural signals. The processing speed is in the order of Mbps, megabits per second, explaining why the sound is heard instantaneously. Simply marvelous!
There's wonder in everything we see, sense, perceive, and experience!

Human perception of sound depends on various factors such as pitch frequency, volume levels, tonality (A sound of specific pitch, listening quality, timbre), range, reverberations, noise, echo, and other variables. It differs from person to person. We hear all the components of the sound as a whole *sum* in totality, and simultaneously. Another major aspect of hearing is '*sound localization*' as to where the sound coming from. A sound is produced in a geometric setting of three-dimensional space.

Therefore, three values or measurement parameters (L+W+D and /or angular coordinates) are required to determine the true location of the source. According to "*psychoacoustics*' the initial step is carried out by the periphery which informs the brain about direction through '*head related transfer function*'. This is the property of *binaural* hearing in which the sound signal arrives at right & left ears with infinitesimal, but perceptible differences recognized by the smart sensory components in the organ in terms of *time delay* and *sound levels*. '*Localization*' (the ability to determine the origin /location of a sound with respect to distance and direction) of sound is mainly performed in the inner ear by *cochlea* by using very sophisticated processing in frequency & '*phase*' (roughly, the rise & fall of the sound waves) detection, analyzation, filtering, and computations of several physical parameters of distance, elevation, azimuth, and other variants such as time delay, sound levels, etc. Some of these results are extremely useful also for '*audio discrimination*' to differentiate between closely resembling sounds, words, and other sound bytes (mainly a broadcast jargon for a 'snippet of recorded speech') we hear.

Psychoacoustics

Psychoacoustics refers to human behavior, perception of sound and response to hearing. The human hearing range is subjective and limited within a range, expressed in units of *Hz*. The sounds that we hear, speak, sing, and produce in an open or controlled environment occur within this range. Various attributes and variables of sound affect human hearing perception such as frequency, sound pressure level (sense of loudness), medium in which the sound is produced & propagated, acoustics of the environment, etc. Pitch is measured in Hertz **(Hz)**; the loudness in decibels **(dB)**. Humans distinguish sounds also by other characteristics by its sweetness, harshness, timbre, and how it develops in its duration of existence as steady, rising, fading, oscillating, bleating, and numerous other distinguishing characteristics. *Cadence* or roughly the speaking style is also a major distinguishing factor. In music, high-frequency tones are classified as *high pitch*, while low-frequency tones as *low pitch*. The musical instruments can produce sounds within and beyond *human voice range* with its own typical timbre, loudness, lushness, resonance, sustain, reverberation, and other individual qualities that humans cannot. These properties help us to distinguish between different instrumental sounds. These factors are very important in the production, deliverance, hearing, learning, and the enjoyment of music.

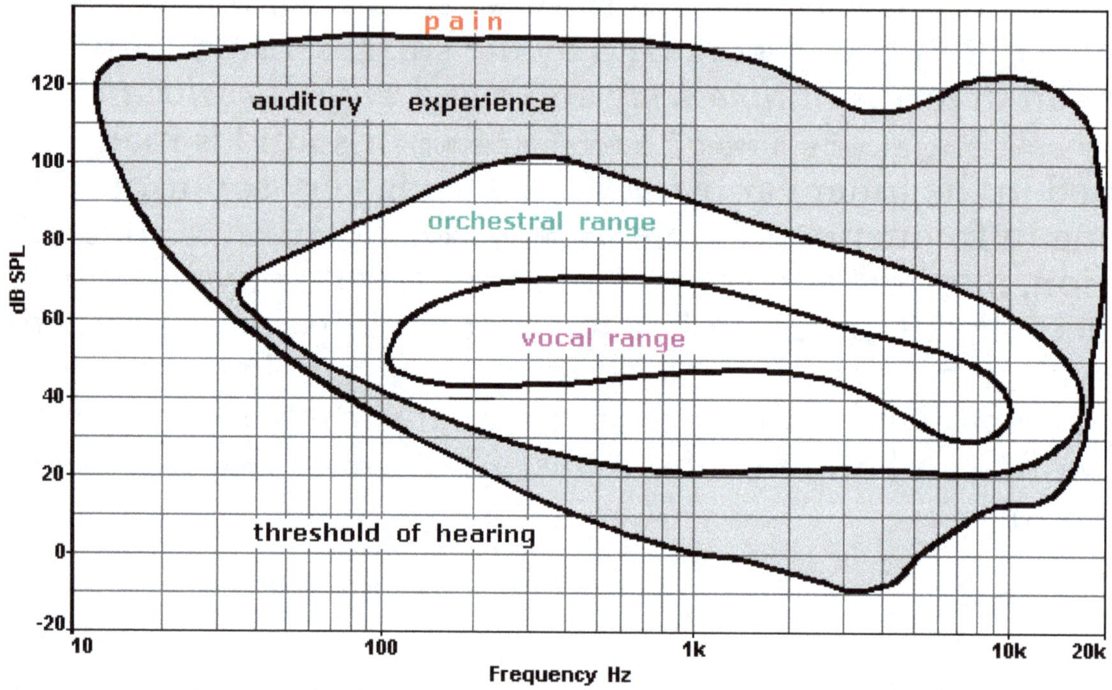

Human hearing sensitivity & characteristics on a *logarithmic* chart

Ears can capture and process more information within itself than any other sensory organs. In other words, beyond perceiving the sound, they perform several functions before sending out the informational data to the brains!

The above laboratory observational chart shows:

- The human hearing range has a threshold of a rather narrow spectrum
- Ability to hear is dependent on both the pitch & loudness of the sound.
- The '*sensitivity*' of human hearing perception is non-linear
- The higher pitched musical notes feel disproportionately louder compared to its Sound Pressure Level (SPL)
- The '*sensitivity*' of human hearing becomes steadily stronger passing the 50 Hz mark

- Human hearing is most sensitive between 1200 & 4000 Hz that humans cannot produce (The human vocal range is typically between 80 Hz/ note E_2, and 1100 Hz/ note C_6)
- The '*sensitivity*' decreases steeply thereafter beyond 4000 Hz
- Signals beyond 20 **k**Hz cannot be perceived, therefore not heard!
- Also, signals below 20 Hz cannot be perceived, nor heard!
- Loudness below -10 dB level is below the human '*threshold of hearing'* and cannot be perceived, nor heard!
- Loudness beyond 120 dB level breaches the '*threshold of pain*', extremely damaging to the ear!

Here are a few other noteworthy facts about hearing:

- Commonly, the hearing range shrinks with aging
- The deficiency is initially felt at the upper end of the hearing range (high pitches above 12 kHz)
- Audiologists use '*audiogram*' (identical to the above chart) to determine *hearing loss* pinpointing the frequencies those <u>can be heard</u>, and at what volume (dB) levels; outside of which becomes hard of hearing.
- Some noises we hear in the environment and everyday life are painful and harmful. Rocket launch, '*sonic boom*' (a loud explosive noise caused by the *shock waves* from an aircraft or projectiles traveling faster than the speed of sound), canon /artillery firing, fireworks, blaring ambulance horn, Hammer drills, jackhammers, motorcycles, & 2-cycle engine noises, rock music, city traffic, gunshot, etc. (in the order of the worst to worse) are such sound levels that can be dangerous to human health and wellness.

- Some strong sounds below the lowest frequency hearing range may be felt on the body as vibrations, though cannot be heard
- Dogs have wider hearing range, up to 45 kHz! The dog whistle is higher pitched above human hearing (ultrasonic), not annoying its *superior* masters!
- Cats have even better; useful to catch mice!
- Elephants have hearing ability below that of the human hearing range of 20 Hz (subsonic)
- Dolphins have a hearing range of up-to 145 kHz, that too with no ears. They only have *ear holes*!

The central nervous system has an important role in deciphering and interpreting sounds entering the ears. It has vast resources of speed, and processing power, learning, and memory capacity. Finally, it presents the information in a *perceptual* way that the sound is most enjoyable to listen to; intelligible to hear, and understand. It is essential to ensure a higher quality sound exists under all circumstances where humans communicate: at home, in rooms, lecture / meeting halls, arenas, auditoriums, concert hall, worship centers, etc. Hearing and listening are crucial skills closely linked to learning. The fact dictates that good acoustics in classrooms is absolutely essential. Nowadays, continuing education of adults and senior citizens are quite common in careers and social gatherings. So, ensuring an improved sound environment in such places must also be a major requirement.

The considerations in music concert halls are only much more demanding. It's a place meant for sweet and supreme quality of sound. There a variety of sounds are produced, sung and played with very extended frequency range, dynamics, tempo, with resonance, and reverberations. Invariably, all sounds produce higher order harmonics called '*overtones'* which are essentially multiples of the *'fundamental note frequency'* caused by the initial vibration. All sounds need to be conveyed to the listeners in its original form, true fidelity, and with the highest '*intelligibility*' (understandability of all speech sounds) without noise, echo, feedback, or other annoying artifacts (corruption of signal by external causes). Physical objects in the paths of sound waves can reflect, absorb, delay, and distort the original sound causing spurious (not genuine or real) interferences.

As this Art Center facility is intended for multiple uses, a few other factors and requirements have to be implemented. Normal speech converts airflow into energy. In speech, we use consonants, vowels, and syllables. They are actually *sounds* and not to be mistaken as simply alphabetic letters like: k, p, s,... Vs. a, e, i, o, u. Consonants are spoken with mouth fairly closed, and require reasonably *precise articulation*. Producing consonant sounds requires constraining air flow at different locations of the mouth and positioning of the lips in certain ways.

On the other hand, vowels are spoken with mouth fairly closed. They are more sonorous (sounding musical) than the consonants. Then, there are '*approximants*' which are a set of consonants that sound somewhat similar to vowels (y, w, r...) and therefore often confused as vowels.
Most syllables contain a vowel!
All of these sounds show its own spectrogram (graphic illustration of frequency components contained in the sound on a display screen) characteristics.
Vowels vary in pitch. Higher vowels [*i*, and *u*], tend to have a higher fundamental frequency than low vowels [*a*]. Vowels are distinct from one another by their acoustic form or spectral graphics. Spectral properties are the speech-sound's *fundamental frequency* plus its '*formants*'. They contain a band of other naturally accompanying overtone frequencies. In rare cases, the overtones may have higher intensity as well. In such instances, they become the major determinants on the quality and characteristic of the sound!

Typically, counting only the fundamental frequency, the whole of human vocal range is sparsely narrow, from 80-1100 Hz. It means that the maximum vocal music ranges only from E2-C6 on a Keyboard. In spite of that, the earlier explanations show that much more features have to be provided to ensure a Hi-Fi sound quality. By nature, a person has only a specific scope in the vocal ranges of *bass, baritone, alto, tenor, or soprano* classifications in the order of low – high. The average vocal range for males lies in the Bass – Tenor range. Musicians who practice rigorously may reach 1.7-2.4 octaves spanning more than one classification; while others just have 1.4 to 2 octave range. The average vocal range for a females fall in *mezzo-soprano* range with 2 octaves. Most soprano singers have the highest human voice range with 2-3 octaves. The common attribute of sound is that '*all sounds produce overtones*'. On an audio spectrogram, it shows crossing several octaves, progressively diminishing in amplitude levels.

We must acquire some basic understanding as to how our vocal mechanisms operate. After all, we use it all the time, even while sleeping! The facial frontal cavities, the nasal and oral cavities extend from the back of the mouth to the *pharynx* (throat). Throat begins at the back of the mouth and up to the upper part of the stomach. The *phalanx*, a tube located at the back of our throats that bifurcates into *esophagus* (the food duct), and the *trachea* (the air duct). The esophagus transports food to the stomach while the trachea passes air to the lungs, the two most important necessities of life! We need to talk and communicate too. Our voice box, called the *larynx* houses *glottis*, the vocal apparatus of the larynx. It consists of a pair of '*slitted*' mucous membranes called *vocal cords* stretched across horizontally with a slight slit of narrow opening in the middle. They **vibrate**, modulating the flow of air being expelled from the lungs. During *phonation* (as we speak), *air is forced out of the lungs*; passes through the narrow gap allowing it to vibrate (something like a flexible, thin loose flap of feathery object fluttering in the wind) making changes in air pressure, producing voice. Various muscles control the movement of the vocal cords. The vocal cords are *open* when *breathing*; *vibrate* when *speaking*. The intricate and meticulously grown mechanism is just amazing and mind-boggling. In lengthy, tedious, and investigative processes, the professionals in the respective fields conduct research and make clear sense out of it. They look, probe, poke, stimulate, photograph, utilize various imaging tools & technologies; document, and record; perform studies necessary to discover each anatomy; sculpt models; conduct detailed examination, and analysis of the relative motions and functions; make inferences, and prove the theory until corroborated. We must salute and commend all the enlightened professionals who persevere to fathom the mysteries of the body and mind.

There are striking similarities between all forms of energy. They consist of changing vibrations of various frequency /wavelength and intensity. Most are electromagnetic radiations. However, audio is not! It's an exception. It's just *waves of air pressure*. All signals need a medium through which it can propagate its energy. One important aspect of sound is that: '*all sounds are composed of multiple vibrations or frequencies*', although our ears generally perceive it as just one combined '*sound information*' much the same way that white light is made up of a wide spectrum of colors. The most visible and recognizable light effect that illustrates the phenomenon is the rainbow made up of several colors.

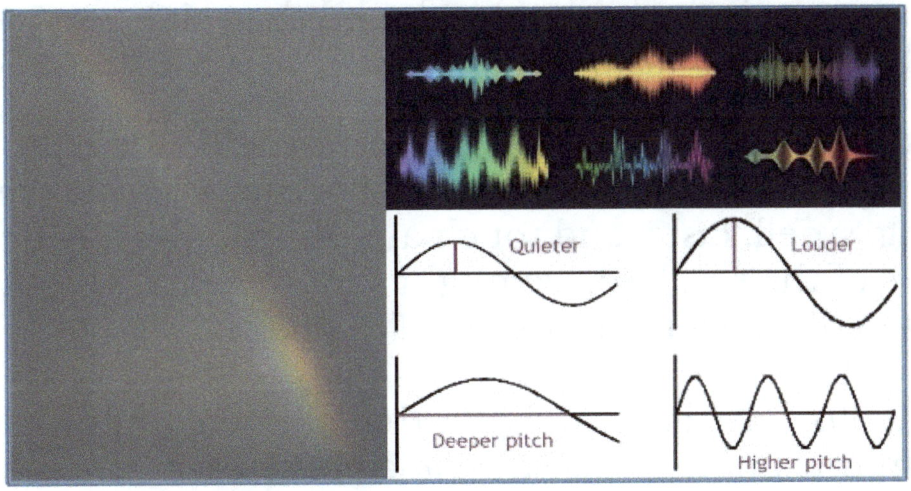

Light *Sound*

The audio frequency spectrum refers to the number of times when a sound wave repeats per second. Frequency is directly proportional to pitch; the greater the frequency, the higher the perceived pitch.

Therefore, one might say: "*the overtones /harmonics are the colors of sound*!" Once again stated, these overtones (frequencies occurring above the 'fundamental frequency') are higher multiples of the fundamental frequency. You may rightfully consider the fundamental frequency as the first vibration produced in the air by a sounding body when energized by an external force. It's also the largest in amplitude. But in fact, associated with the fundamental, there are a number of harmonics, and also subharmonics (fractions of the fundamental frequency. Thus, note A4 = 440 Hz; A3=220; & A2=110; A3 & A are subharmonics of A4 by fractions of 2 &4), a few of which may be strong partials resembling **harmonics**. [***harmonic*** in physics are component frequencies of the signal of a wave that is an **integer multiple** of the **fundamental frequency,** while **partials** in mathematics are **partial derivatives** of **algebraic function**]. The timber varies depending on the number of harmonic (overtone) frequencies, and their relative intensities contained in sound. Therefore, it is the *signature ID* in differentiating between sounds from one to another. Reiterating the same fact, the abundance, and types (harmonic, in-harmonic, odd, even harmonics & the varying intensities of each group; the mathematically precise proportions of the overtone scale) of the overtones in the spectrum of sound help us to tell the difference between one musical instrument from one another; the same way, one person's voice from another. It's exactly how a newborn baby initially recognizes its mother, yes, by her voice! Second only to touch (tactile), the hearing (auditory) sense is the better developed than sight (visual) or the others, smell (Olfactory) and taste (Gustatory) senses.

So, talking to babies is extremely important! The nature makes them smart right from the start! The five primary senses we possess—sight, hearing, smell, taste, and touch are philosophically described as "the windows of the soul'. So, we might say "we learn, discover, and enjoy things in life through our senses!" *Fantastic*!

In an auditorium, it necessitates that all the components of a good sound system must have *sufficient bandwidth* to carry the full range and elements of the sound signal.

In music and orchestra, the harmonics play an important role. Higher-frequency waves occur with varying prominence and give each instrument its characteristic tonal quality giving the instrument a dominant role for that musical segment. That's one of the primary reasons why the conductors select suitable instruments at various segments of the music. A fundamental tone alone sounds unnatural or like an electronically produced "*synthetic* sound by a *tone oscillator*". Well, at the same time, it's to be pointed out that too much spread in overtones muddle the sound; adversely affect the intelligibility, diction, and quality of the sound.

The sound characteristics of various music instruments used in orchestra are elaborately discussed elsewhere in the book. In any instruments sound emanates from different parts of the same instrument; all tones sound together as a whole. The orchestras use an array of instruments in the so-called '*families*' of strings, brass, woodwind, and percussion. Each instrumentalist sits and plays in the same group. On the stage, each group is seated in a traditional system of seating arrangement. It's very helpful in radio broadcasts of orchestral music, as listeners can detect precisely the source and direction of all sounds. The sum total of the sounds the audience hear feels louder and expansive than a single one. Besides, the sounds '*beat*' together and produce even more overtones and combinations of them.

The lead vocalist and instrumentalist are always in front of all the others. This has a major visual and aural impact, as their sound waves from the front will reach the audience earlier before the others, and will therefore be perceived and registered by the brains as the '*first in line*' (remember the *Que lines* at ticket counters or bus stops!); sounds prominently than the rest. The mechanics of the cochlea in the inner ear permit the initial sound waves to have a remarkable and profound impact on what is actually heard. The transfer rates at which the data is converted often vary; some data packets may get lost in its transfer from the outer ear to the auditory nerve; and/ or from the auditory nerve on its way to the brain. In summary, sensory perception involves the organization, identification, and interpretation by the brains of the signals received via the receptors and partly processed by the sensory organs themselves. The brain sends out its meaning as impulses to the mind and body as the case may be, to feel it. Naturally, the better the data, better the results.

It's now an opportune time to discuss the usual audio controls we fiddle with on the electronic sound systems we have, to suit our listening taste and preference.

Each voice falls in its particular, natural pitch/frequency range possessing unique musical qualities. The bass range refers to the lowest voice range roughly from 40 Hz to 400 Hz. Bass guitar, double bass, Tuba, Mexican Guitarron, Bassoon, certain drums, etc. sound in this range. The bass are often recognized by the bigger physical size of the instrument, as the wavelength of the sound they produce is longer (but lower in frequency) and hence the deeper sound! The bass offer the foundation of rhythm with which the music can move forward. It jives with our bodily rhythms-pulse, walk, jerks & movements; sets energy levels as the strong sound vibrations can be felt all over the body. The *low midrange* is sometimes referred to as the '*bass presence range*'. It contains all the low order subharmonics of most instruments. Every time music lacks depth in its instruments, the problem is directly associated itself with the lower midrange section.

The mid-band is in the *sublime spot* of listening pleasure. Most of the contents of communication reside in this range consisting of the melody of the music, verbal communications of all kinds including film, TV, & drama dialogue, sounds from the surrounding nature, synthetically generated sound effects, etc. The tones in this range sound lush, warm, and natural. The most part of the melody dwells in this part and reaps the greatest accommodation and hence the joy of listening. Too much sound level is the sole culprit that can make the sound harsh and spoil the enjoyment. The high band contains overtones with unique attributes; provides timber and '*brightness*' to the sound, providing sound identification and help in localization of sound.

We use '*equalizer*' control knobs on home entertainment system to amplify sound and adjust responses of certain family of instruments we like hearing or falling in a particular voice range (low, middle, High, correspondingly labeled as bass, mid, & Treble / high). Thus, the entire rhythm making instruments or specifically all the instruments in the *Bass / F* key may be treated as '*bass*'; string and most wood instruments as '*mid*' in frequency range (382-1.4 kHz.) Piccolo and brass as 'treble /*high*' (from 4 kHz-6 kHz). Treble provides clarity and definition to sound. The treble control can improve sound quality significantly and provide a kind of *immersive* experience. Upper Treble range of 6 k to 20 kHz is referred to as the '*brilliance range*' as it enhances Hi-Fi quality. Finally, when the frequency balance is ideally set, we feel that our listening experience has climaxed!

The lighting design gives utmost consideration for energy efficiency, use of sustainable and renewable energy independence. The stage lighting requirements are vastly different from the conventional ones where harsh, intense light is cast on to the stage. This causes a lot of discomfort and harm to the performers. The lights cause skin burns, blemishes, heat, sweat, glare, and shadows among the many other serious health hazards.

The newer cameras use highly light sensitive solid state image sensors for image capture; produce negligible noise and artifacts in the pictures, which if any, can be removed by filters and Digital signal processing. The lighting is designed to have a 3D illumination on the stage missing no act or expression of the performers, nor shadows obscuring. Both inside and outside the hall, bright LED lights are used for all lighting requirements that produce negligible heat. They are dimmable and can be adjusted uniformly, and are chosen to faithfully simulate various lighting scenes- daylight, cloudy, partially cloudy, vastly cloudy; foggy, starry, moon-lit nights, etc. The ambient temperature has significant impact on sound propagation adding to the lighting task another degree of complexity.

The main stage lighting is from a large ring suspended from the top. They cast ample light downwards and has light arrays pointed all around to illuminate the faces of performers, and instruments. The inclinations and the intensity of lights can be individually adjusted precisely and remotely as the scene demands. The lights are voltage / color-sensitive in the sense that they can produce various tints of colors by changing the voltage values accordingly. During a performance, the light system operators show off different '*dancing & flashing*' light effects to the thrill of viewers. They are especially useful in theater performance and special effects, besides the wild and hallucinating rock concerts!

In addition, there are spotlights, which are comparatively intense lights installed at strategic points; some from the balcony positions shine onto the stage; some on the back walls, and others from the side walls. Each performer on the stage can be individually selected, highlighted and tracked, or followed with adjustable '*beam widths*'. The opera performers, comedy show artists, and animated music performers move around the stage a lot. They don't want to be seen as momentarily disappearing from the stage!

Another innovative design is to surround each camera with a '*ring of lights*', which, up on preview, turns on the lights onto the subject in a subtle manner. Most '*lighting*' devices are mounted on P/T revolving heads allowing it to be '*remote controlled*' with large degree of motion and adjustment. That's how spotlighting of a specific object or location is usually achieved.

The entire Art Center complex is a '*high-tech*' centric facility, powered by electronics, computer circuits, and green energy technologies. No payment or transaction takes place at the premises. If assistance is ever needed, there is a large well-trained, groomed, educated staff, always ready to offer support, and services.

Following the requirements and guidelines of the "ESG" (Environmental, Social, and corporate Governance), it is branded as a '*beyond zero emission*' structure; using only green, renewable energy for its entire needs. It has its own green power plant, with an array of solar panels, tucked on the top of the roof of the auxiliary buildings. They are well concealed and blended in to the surroundings. If at all anyone can spot from an elevation or distance, it is seen as just reflecting the cloud and the sky above. In addition, there is a continuous line of solar panel arrays along the perimeter of the complex, so positioned that the shadows cast from the line of trees are minimal. Rainwater is drained from the roof into many subterranean cisterns / silos lined up on the outer periphery. The water is further purified and oxygenated to remove microbes and virus at a water purification plant run by the municipality. The purified water find its way to the many hi-rise flats springing up around the city. Any unused amount is used for irrigation purposes in parks and gardens.

The parking lots have sunroofs with solar panels. There are 150 designated '*reserved & paid-for*' lots for electric cars and 20 Hydrogen fueling stations. Each charging station can charge 4 vehicles at a time in a cyclic '*pulse charging*' mode. Practically, 3 modes of charging are available. They are self-sustaining drawing no current from the public utility grid. The smart charging stations can recognize each EV vehicle and its charging requirements as well as the plug configurations [not all of them are SAE 'Soceity of Automotive Engineers' standard compliant]. They are capable to accept slow, or DC fast charging, and /or autonomous charging. All the current charging parameters are displayed on the display panel and safely shuts off at 80% of full charge capacity, avoiding any potential overcharging, or fire hazards. To begin with, the charging station runs diagnostic checks for '*short circuit*' in any of the individual battery in the battery modules. Then, it downloads the critical technical specifications of the battery packs: Battery state of charge (how much battery capacity remains in the battery), Power (kW), Voltage (V), and current (Ampere-hour) rating, Battery charge/discharge regimes, etc. Some types of batteries need to be fully discharged before charging. Some others shall never be fully discharged! Typically, charging stations are *Electric Vehicles* (EV) *specific*. When all safety conditions are met, the charging can begin. During charging operation, the station monitors the instantaneous battery temperature; promptly, shuts it off if any measurement exceeds the safe operating limits, or at the end of charging cycle of 80% full battery capacity. Each outlet can fully charge a typical 90 Kw battery module in just under 2 hours. That is the estimated average specified time for an event. A '*Green Hydrogen*' plant is under construction. It is mainly intended for additional /redundant/ future power needs as well as to provide Hydrogen fuel for the so powered future cars. A relatively new technology of solar hydrogen production from Hyper Solar, is seriously contemplated. It utilizes only abundant and easily available natural materials; adopts the natural principle of photosynthesis (Photosynthesis is the process by which plants harness energy from sunlight to split water molecules, producing oxygen and carbohydrates) used by plants to manufacture its own food. In this process, the system utilizes specific catalyst material, solar power, and techniques of '*atomic layer deposition*' on the 2 electrodes to optimize water electrolysis mimicking photosynthesis in separating hydrogen from water. The hydrogen flow, in turn, is used to drive fuel cells producing electricity. This technology simply needs any source of water, even waste water, sunlight, and a low-cost catalyst (significantly **not** the usual kind of heavy '*precious metals*' Gold, silver, platinum, palladium, etc.) immersed in water to produce environmentally friendly green hydrogen. The byproduct is H_2O, pure, potable water. Hydrogen is the most abundant element and the cleanest fuel in the universe. The hydrogen produced this way is purely clean, unlike the so-called gray / blue hydrogen, which is commonly produced by '*re-forming CO_2 emitting*' natural gas (CH_4). Fuel cells combine H_2 and the pumped-in O_2 from atmospheric air to produce electric energy. They are compact and clean; the byproduct is again pure water. "*Water produces water*" by regeneration! The technology is amazing!

It is widely known that the water given out from the fuel cells on the orbiting space station laboratory is the only available water source on it. It is used for drinking and washing purposes by the astronauts. Yes, it's potable! (drinkable). What a wonderful natural element is Hydrogen that is also abundantly available in the universe! It can only be more wonderful if it were edible too! Then people won't starve so much; H_2 will be the most humanitarian element, in addition to its other fabulous properties!

In the Vedic literature "*Pancha mahabhutha*" (the five great classical elements of the universe namely: Bhoomi or earth; Jalam or water; Agni or fire; Vayoo or air; Shoonya or vacuum/space) states that all creations are made up of these classical elements. Sun should have been at least '*honorably mentioned*' in the category for its supply of light, heat, and energy. It may have been left out, due to the fact that it is not an element in the creation, although it is essential to maintain every other one of them. Enormous power and capabilities reside within these elements. The technology is finally inventing ways to separate them from other compounds, and putting them to everyday energy uses!

There is no power lines running through the compound. Parking is always free (except for charging station lots, discouraging other cars from taking up the service lots); the service charges for electric charging and Hydrogen fueling is for covering all the costs of generation and maintenance only. It is currently not meant as an additional revenue source. This is not a capitalistic enterprise, purely socialistic! In capitalist countries, it's corporate welfare; in contrast, here it's public welfare!
The identical policy is applied for use of the facility for any charitable and service organizations depending on availability. It is meant for cosmopolitan access and usage; for noble reasons. Besides, major international conferences, seminars, and interactive Telemedicine conferences qualify for the facility's use. No political, religious, radical events, or wedding receptions shall be hosted at this venue.
It is aptly described as an "Arts Center" and hopefully will remain so forever!

Once registered, the tickets can be received on smartphones. It is all machine read, scanned at service locations, through other secondary, or tertiary options. Everything is billed on the information collected at the issuance of ticket. The seating is reserved with the purchase of tickets as in the airline bookings. Each seat is graded by its rank in visibility, sound reception, comfort & convenience qualifications. The entry to the hall is managed by identifying and processing the 'QR / barcode', read directly from the digital tickets.

The surrounding environment presents a friendly, affable, welcoming character. At night, the dome shines in the dazzling laser light display of variegated (displaying differently colored spots, patches, streaks, etc) beams, providing a brilliant ambiance, highlighting and glamorizing the dome at the geometric center, which shines like a tiny distant galaxy from the International Space Station. The air traffic route deviates only by a couple of Kms from the location, sparing the nuisance of air turbulence and engine noise. But, it is still visible to the passengers, offering an enigmatic, and impressive aerial spectacle, as well as a friendly welcome up on landing.

The internal settings marvel in splendor and technology as well, rich with audiovisual gadgets, displays, embossed paintings, and carvings. Most of the devices are embedded. There are many high definition cameras on stage; on the three layer balconies; suspended; PTZ (Pan / Tilt / Zoom) & robotic cameras, covering everything, every location, the musical instruments, the performers, the choir; the chorus singers, the audience, and even the dome. The cameras are compact due to the inherent nature of solid state technology. The lenses for the cameras do not need to be large, bulky, and cumbersome. The cameras perform its function as image collectors; zoom, scaling, and the essentials are accomplished by Digital Signal Processing by custom-made '*SoC*' programmable chips. SoC chips encapsulate an array of electronic circuits on a microchip consisting mainly of CPU, GPU, 3D flash memory technology, that has 176 layers of memory cells stacked on top of each other, and Input/output (I/O) interfaces for external interface.

In its physical shape, each tier of the balcony surrounds progressively closer to the stage from both sides in a perfectly symmetric fashion with respect to the middle isle. They are cantilever suspended structures, invisibly anchored at one end, appearing to be giant vessels floating in the air. The belly of the balconies are hand painted with authentic, original, dusty colors popularly seen ubiquitously on Indian monuments. The safety rails are studded with polished wood in its actual tint, using no stain, or veneer. The Sheikh like to use symbolisms in most everything; makes it a '*teaching*' tool and a useful lesson. He is particularly fascinated by the Mughal architecture, paintings, historical facts of conquests, annexation, subjugation, and expansion of the empire. Each of the first sons among his 23 children, ('representative of each of his chromosome') from his 4 wives is named after a Mughal emperor "Shahjahan" from the first; Jahang for "Jahanghir" from the second; Bahadur from the third, and "Akbar" from the fourth, and the youngest. Invariably, the youngest has been always the most favorite of all, whether in the case of wives, or children. Significantly, the Mughal emperor Akbar was coronated as a teenager, at the tender age of 13! The Mughal dynasty with *Persian / Tukish / Mongol & the Caucus* heritage and from the Central Asian region, invaded India in the 16th Century and ruled vast parts of Northern India for nearly 300 years. They spread their Islamic faith, culture, and skills all along their routes to India, (Afganisthan, Pakisthan, Bangla Desh). Unlike the modern day '*caliphate*', the Mughals were relatively tolerant and tried to maintain communal harmony with the majority Hindus. Their contributions and cultural influences are still clearly visible in the monuments & mausoleums they built, the Urdu (Persian script with a combination of mostly Hindi words + borrowed words from Persian & Arabic) language many still speak, and in the Hindusthani music which is distinctly different from the South Indian Karnatic music that virtually remained unaltered for ages. Even in the present day, many World political decisions and policies are strongly influenced by the traditional, historic ties, and relationships between these countries, whether they are energy embargo, cultural exchanges, trade policies, or Uranium enrichment!

In the auditorium, the seats are cozy and functional, made by the automotive/airline seat maker Adient. There are '*Hi-res. Infotainment*' (Television term for presenting factual matters & data in entertaining means) flat panel screens as in automotive dashboard and on airline seats. The audience can switch to various input sources such as video, Lyrics / Music sheet of the song, auditorium centric acoustic, and environmental data, such as echo, reverberation, sound volume, audio signal waveform, etc. Internet browsing is also enabled. The viewers can also switch to various cameras by location, or camera angles to see the most desired action on their own assigned monitor. The technology is supported by fast "SoC" (System on a Chip; much similar to the heart pacemaker / defibrillator for regulating heart rhythms which integrates battery, sensors, signal generator, timing, and inpuut terminals) and microcontrollers (MCUs) / from Cypress semiconductors. At the end of the show, the seat- mounted '*tablet display*' devices are automatically uploaded with the video recording. The viewers can swipe their PDA (personal Digital Asst.-smartphones) over it to download the entire program instantly in contactless data transfer (Near field communication 'NFC') mode. The audience are spared of the trouble of initiating and constantly monitoring their own recordings.

Almost all equipment have embedded and programmable semiconductor modules with '*edge servers*' equipped with 3D NAND micro-memory chips from Micron Technology, which can hold large volume of data in vertically stacked layers. Then, there are '*Qi-standard*' based fast wireless charging pads at each seat, thanks to Integrated Device Technology which allows both Android and iOS smartphone charging. The isles and pathways are layered with pressure sensitive luminescent material. They illuminate with just adequate diffused light leading by an arrow to the reserved seat or to the '*Exit*' as the case may be. They seem to be knowing where a person has to go! It's a '*virtual usher*'!

The seats are equipped with proximity and motion control systems; incorporate seating solutions that can be easily modified to suit many comfort levels. The built-in 'AI' assess body position and pressure points on sitting, leaning etc; protects '*piriformis*' tendon and *sciatic nerve* (*Med.* hip muscle, back, buttocks, and the big nerve passing through them). Periodically, gives a gentle soothing massage, helping blood circulation; helps people to enjoy the performance without feeling any strain, stress, or after effects. The audience is pampered, served, and accommodated to unprecedented levels of luxury, comfort, and convenience. It may be too much to accept for the simple-minded people with a philosophical and minimalistic mindset. But, that may not be the case with the mainstream world of insatiable wants and demands of luxury!

The truth is that it may very well be the future standards of normal living, designed to mitigate the ill-effects of aches & pains, stresses & strains caused by the hectic lifestyle of the modern work-life culture!

Using the interactive screens, the audience/ viewers are provided with facility to pre-order beverages, snacks or any amenities at the intermission; also send urgent alerts for medical emergencies.

Each seat is equipped for immersive aural and visual experience. For this dual purposes, a display tablet and a noise-canceling headphone are provided. By default, the headphone switches to corresponding selected signal on the display. If a solo part is played, the display panel by default switch to a full screen picture of the soloist and the solo instrument. The solo music is heard through the headphone. The specially designed headphones use 8 *super cardioid* microphones with extreme directionality to augment sound location. The "*Learn*' function on the tablet shows all the location of these microphones; 2 L /R overhead, one each Front- L, Front- RL, Front-M, and similar setup for the rear to pick up weaker reflected sound. The size, posture, and shape of each person and his /her hearing situations are different. Therefore, the noise-canceling features processed are customized and optimized based on these factors as well as the environmental conditions of the halls. The headphones utilize very sophisticated noise reduction algorithm developed with AI machine learning using multitudes of voice samples to suppress ambient noise resulting in a dramatic reduction in high frequency noise. With Precise Voice Pickup Technology and advanced audio signal processing, the sound become so immersive and so real. The listeners enjoy true 3D Reality Audio!

The opening ceremony and the inauguration was conducted in great celebrations, media blitz, with all customary '*pomp and circumstances*' (celebration accompanied by traditional formalities and ceremony), laser light shows in the presence of celebrities, the pesky political figures, government officials, and invited public who earned distinction in their specialized field. The chief guest, the governor of the state HE. Shri. Bharath Bhushan who was accompanied by the retinue of the chief minister, and his ministers of the cabinet did the honor of inaugurating the Art center. This project was good for ministers and government officials too. They were enriched by lavish sums of ill-gotten wealth at each step of the way, on every permit they issued, without missing, or mis-counting a single one! In addition, their blood relatives received top jobs.

Anything happening in the land is considered coming out of their bounty and benefaction, and therefore justifiably deserve rewards, the bribes interpreted as bonuses. The deed of corruption & bribery with dubious moral underpinning is routine practice in politics!

One has to be a consummate scoundrel to be a minister in the cabinet. There is a lot of competition for every position. This is sadly the widely resented '*standard order of political ethics & morality*', in the whole world right now, from superpowers to third world regimes, and banana republics. It's etched into the consciousness of the public so that they accept it as the normal administrative procedure and protocol. They don't want to fret over every thing they see. They have a life to live; many things of immediate and peremptory concerns on their minds. Citizens could hardly be bothered and distracted!

Among the attendees, were many aspiring artists, school children, students in fine & performing arts, prominent institutional, business, and industry figures. The public was given entry to the sprawling complex. They are provided with informational flyers, posters, hats, goodies for children; refreshments, and amenities. The next three days, the public are given official tours of both the hall, and the complex in different groups. The logistics and the execution of this was done in a methodical, organized, and efficient manner. When one group is coming out of a facility, another one is going inside. At the end of the tour, the guests are escorted to the parking lots to the tour buses, or personal vehicles in a continuous, orderly movement of people. That, in itself is an unusual '*cultural change*'!

Two full days are reserved for stage performers. They are given orientation and pamphlets containing details of the facilities, the equipment, instruments they have to use, performance procedures, the Television / Video/ audio/ lighting resources used in live, recording, and broadcast operations. The initial protocols are now merely guidelines, which will be continuously revised with additional practical knowledge; as and when the installed technology is updated. The aspirants are shown the exact positions, and locations from which they are expected to perform. Many documentary video segments were made. They are posted and available on the website for self educating oneself.

Being the first of its kind in India, there is great excitement and expectations about the entire complex, especially among the participants in the performance world. There are several case studies written about it detailing its features and the technologies employed. They range from design considerations, design models, and construction of each section. They have done several animated sequences and studies of its practical operations using models of audience, performers, instrumentalists, and the choir in a live '*run thru*'. They used dummies fitted with microphones in the ears to study the sound signal quality; cameras on robots for analyzing views from selected spots. Beyond that, the light situations (view, shadow, beam width, etc. as well as the proper camera positions; pan & zoom situations) were tested to determine optimum results. Various entities made documentaries and wrote rave reviews extolling the many remarkable aspects, technical innovations, and the enormous impact it will have on the resurgence of the city as a center of culture, and arts. Several glowing comparisons were made with '*DR Koncerthuset*' in Copenhagen, Denmark for its exceptional sound reproduction and '*Wiener Musikverein*' in Vienna, Austria for its splendor, and so on. In general, the writings generated a great deal of interest, curiosity, and expectations. People traveled from all parts of India to the complex to experience the described marvel in firsthand, face to face evaluation, and assessment. The nature scientists and environmentalists are also excited about this venture. They flock to the spot in the style of fervent pilgrims. The corporate executives sent '*business development*' strategists to study business prospects in these applied technologies. Some heavily pressed fossil fuel industrialists sent 'ESG' personnel to learn and implement similar '*carbon reduction*' technologies in their own operations in an effort to avert paying "*carbon tax*", as a penalty for polluting.

For extensive self training purposes, and listener-experience, the Art center use a product called "*Uncontained VR Attraction*" from **XR Immersive Tech**. The 3D '*metaverse*' technology uses virtual reality *VR* and augmented *AR* headsets. Everything in metaverse world is virtual, imaginary and in 3D space. The user/performer is within that space and environment. A replica of the inside of the Art center concert hall is readily available to be populated by selecting '*objects*'. Thumbnails of objects are also available. Additional objects may be virtually created block by block, say an orchestra, podium, etc. One can select various roles for oneself, for example a '*live vocalist*' performing; *order* a pianist; a choir; listeners, and the like or can even create many virtual objects such as background, curtain, flowers, etc. The orchestral instruments automatically appear on the stage in standard locations, when *ordered*. Immediately, all the accessories needed for the performance are furnished from the large machine Learning library of files. ["Machine learning is a branch of AI and computer science which focuses on the use of data and algorithms to imitate the way that humans learn using sensory organs, environmental stimulus, social interaction, etc. gradually improving its accuracy"]. The performer gets a microphone & headphone; the pianist, a piano, the listener a seat of his choice where audiovisual accessories are provided as standard user items. It can also read and play music sheets if they can be opened in certain specified applications. The program follows all the playing instructions, tempo, dynamics, etc.

The general settings of lighting, sound level etc are activated. If all the settings are satisfactory, the '*virtual performance*' can start any time. The audio comes alive as sound is detected. The vocalist may sing and listened to all the performing voices via the headphone in 'live' performance. The selected cameras are switched on; The '*video switcher*' compose various camera images for the '*virtual broadcast*'. When the piano plays a piece in accompaniment, it's highlighted; camera switches to the pianist first; then include the performer/ vocalist in a 2 way split when both are actively performing. It may later transition to a 3 way split to include an expressive listener who enjoys the music so much. The singer may decide to wander a little bit; approach the pianist or turn to the choir, and so on. The beam of light faithfully follow the moving objects; so also the cameras with the right perspectives. The '*user*' is able to control settings, movement, locations, etc. according to needs. The default settings and actions are dictated by the '*Machine AI*' intelligence and all '*the learned, practical experiences*'. They are so remarkably resourced that most users use the default settings as appropriate and satisfactory.

The 'Mela' associated with the opening is scheduled to last a full week. The travel, tourism & hospitality business were looking forward to the event with great anticipation. They are now reaping rich rewards. The results will be shown prominently on their next few "*quarterly reports*" expunging (erase, delete, or strike out) the '*incentive*' bonuses earned solely by the top management. It's a sullen fact, in the corporate world, CEOs earn on the average 440 times more than the average salaried employees! The wages of average workers stay nearly stagnant for the past 15 years or more! The employees carry out their hard work solely by themselves. The CEO, CFO, and the top cadre do almost nothing by hand, except signing; shed no drop of sweat; need a staff of specialized junior executives and often engage '*outside, third party, specialized consultants*' for research & report to assist them in their '*decision-making*' role! They have board of directors to look after their needs and rewards while the workers, the largest work force in any company have none! Many companies go out of their way to stop '*unionization*'! What a wonderful world for a very few and a miserable one for all the rest!

Chapter: 20 Inside the Amphitheater this evening, it's

Elise's concert

Elise is the darling of millions of fans as seen from her social media pages. She is the premier entertainer of the day. All the 20,000 seats are reserved and booked months in advance. Among them, there are over 1000 celebrities who have distinguished themselves in various fields of service, occupation, skills, sports, film, Television, and other highly visible roles, and fields.

All are now waiting for the HRH Sheikh, the main sponsor and benefactor to arrive with his large entourage of close family members.

Quite early, the Sheikh has arrived and relaxing at the reception pavilion at the far end of the complex with his family and retinue. He is as excited as all the others to see his vision fully materialized. He had chartered a jet to bring his entire family of 187 members. They are now lead in musical procession along the 2.2 Km long winding promenade from the pavilion to the Art center. None of them has ever walked that far in their entire life. They have been always driven or flown to places. This is a major sacrifice on each of their part. But again! This is a procession, like of which they haven't ever experienced; most likely never will! As is the custom, 11 elephants adorning large head worn golden ornaments as in temple processions, one behind the other, (the promenade is not broad enough to accommodate 2 of them in a row; they are not 2 'elephant-bellies'-wide!) leads the parade. People lined up on both sides of the route. They are no strangers to processions; regularly take part in such events in many festivals in the land: cordial, social, solemn religious fiestas; impassioned political rallies; civil demonstrations, as well as the most detested, antisocial harthals. This is an ancient Indian custom, known even prior to the time of Plato who said "*nothing is of more benefit to the state than the mutual acquaintance of its citizens*". Here in India, the animals are also included among its citizens! The revelers carry balloons, flowers, banners, and iconic models of the magnificent Art center. When the procession pass by, the waiting crowd join behind it from each side, constantly, swelling in size. The colorful feathery turban wearing parade masters marshal the crowd and organize the formations.

Leading the parade, right in front of the elephant brigade is the Kerala brass & wood band ensemble playing marching tunes. The Arabic band composed of '*Oud*' (string instrument), '*Kaman*' (violin) '*Ney*', (flute instrument made of cane) '*Riq*' (tambourine with 5 sets of pair of brass cymbals around its circular body), and 2 metallic & 2 wooden '*Doumbeks*' (Goblet drum/ Arabic tabla) play the favorite Middle Eastern nomadic folks tunes, walking backwards serenading the Sheikh's family. The Sheikh shows great joy and appreciation for the music; laughs heartily while counting the pearl beads on his *misbaḥah* (prayer beads). Lined up right behind him are his 4 wives dressed in long dress, top to bottom with scarves in different shades and patterns of the desert '*sand dunes*': red, brown, tan, or ivory. They are tall; look beautiful with dark brown and wide eyes; raised nose, broad forehead, dark eyebrows, and pointed eye-lashes.

Notably, Akbar, his wife Nadia and all 4 of their daughters Zarah, 7; Zeinaba, 5; Zubeida, 3; and Zabrina follow next. Zabrina, the toddler, just turned 2, is dressed in rose colored frilled frock, red satin sashes tied in decorative chain links of varied colors of tassels at the two ends swinging as she takes each step. She is also wearing matching bootees, and shoes. Her delightful wide sandy colored eyes glow like thc waxing moon; thc straight and raised nose is typical features of the desert dwellers. On her head, she wears a silver, studded tiara; looks gorgeously cute, and delightfully pretty. She wriggles away from the fold of her sisters, runs to her father and lovingly clings to his hands. She refuses to be carried; wants to walk in her stutter baby steps; draws all the attention and glamor. Heartily, people cheer her on. She stares at them in child's innocent curiosity. Now, the parade can proceed only at her pace. She has become the little '*Miss. Grand parade princess*'!

All grandchildren of the Sheikh follow right behind, then the 22 of his sons, all in traditional Arabic dress. The wives are all modestly but fashionably dressed with head-scarves and hats; those in one row wear sarees; the next Salwar & Kamiz; the one next in Arabic and the last, wear typical office suites. They are all well-educated, fashionable, sophisticated, progressive, belonging to well known aristocratic families in their kingdoms. They are graduates from Cambridge or Oxford; Imperial College, or Edinburgh. The sons are mostly educated in the Ivy League colleges: Princeton, Harvard, NYU, Cornell, or Wharton. Among many things, the Sheikh values education the most!

The Indian flower girls in long ponytail, decorated with flowers approach from either sides and shower them with rose and fragrant jasmine & white peony petals. A group of teenage girls of 4, step forward and anoint each of the little girls with forehead-dots. Zabrina tries in vain to reach for more of it. She thinks it's '*toppings*' or '*icing*' on a cake. Then uniformly dressed young ladies appear and do the same for Nadia and other women entourage; some of them politely decline. Curtsy of the jewelry merchant houses of Trichur & Ernakulam. Another group of ladies, cladded in long sarees of rainbow colors and patterns supplied by Seemati Silk house and ornamented with fashionable necklaces studded with glittering gems and shining bangles crowding on each of the arms from the elbow to the wrist, also wearing dazzling, dangling earrings, display their gilted trays of jewelries in front of the princesses. Some of them touch, handle, wear, and cast around in inquiring looks. Each identifies their favorite item with name initials, on the price tags. Indian fashion jewelries are coveted among the elite of the world. They make jewelries in all categories of personal adornments; all in genuine 22 karat (Twenty-two **karat** gold is the high-quality gold, made of 91.6% pure gold, the highest-graded gold alloys used in jewelry making) gold.

They are made by artisans and goldsmiths with long continuous craftsmanship tradition ever since they migrated to India following the fall of the city of many skyscrapers, and known as the cradle of modern civilization, Babylon, where jewelry making started and had been flourishing since more than 4000 years ago. The goldsmiths are believed to be from the same tribe as "the richest man in Babylonia", a goldsmith himself. The jewelry shown here are exceptional; made by laser welding tool for pinpoint precision, that fuses the pieces together, seamlessly joining them as one integral part with the main body; without blemishes, or marks. The '*encrusting*' of gemstones are carried out by precise inlaying and / or overlaying to show stark contrasts. They stand out and add brilliance. They are exquisite! They cannot get such gorgeous ornaments from any of the gold '*souk*s' in the native country. The princesses feel stylish and glamorous; looks around in great pride and self-awareness.

These pleasant distractions mitigate their strain in walking! The crowd enjoys their presence and unique mannerisms.

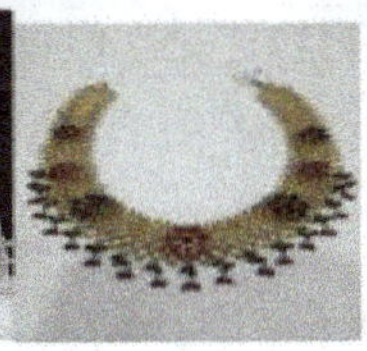

The camera men strain to get a focus on the jewelry items in a tight angle. To the great surprise of the onlookers, none seem to be shocked or perturbed by the hefty price tags on them. Each of the selected items along with many more, will be shown to them later on, at their convenience when they retire to their pavilion after the festive concert. Indian jewelry is very popular in the Gulf countries. They prefer 24k (Karat) gold of the highest purity; but they are too brittle to make intricate and fashionable ornaments. So, everyone, including the super rich, settles for 22k of slightly '*less pure*' which is made into an '*alloy*' by adding copper and/or silver, making it suited to work on. Nevertheless, it still shines bright and invite inquiring looks! The men like Indian spices and fragrance. They are reminded of the *spice trade* their ancestors had undertaken and dominated for long. The women love the glittering yellow metal of fabulous design-works arranged in colorful patterns. There will be a family fashion show where they can wear each of the selected items of jewelry in front of their loved ones.

Naturally, the sales-girls will recommend the priciest item as the most stunning piece on them. The top sales lady may get award and bonus in front of her jealous peers! Furthermore, the chosen jewelry items will be securely brought inside their cabin of the aircraft without any undue worries with the custom checking and clearances at the departure gate. Instead, they will be gently and ceremonially escorted to the leased jet for their departure without any hassle, or embarrassing fuss. No baggage check will be necessary! In India, mainly, the '*custom clearance*' means checking for gold imports or foreign currencies.

Gradually, sun is setting when the slow procession arrives at the front courtyard, where they are greeted with the traditional repertoire of BharathaNatyam dance under the delectable settings. It is followed by an '*aerobatic*' performance by the Kerala Martial Arts troupe. They display various offensive and defensive skills; daring saber rattling; jabbing & evading; attacking & retreating; jumping & ducking moves amazing the fans. There actions are dramatized by sound effects and live instrumental music. The performance can be watched on just '*single-atom-thick*' Graphene-based LED television screens wrapped around the courtyard in the shape of a giant dome. The spectators feel that the saber acrobats jump over them and take a swing at them. They hear the '*swishing*' of the sword while it slashes the air; the clang of the sword when defended by the shield; the landing of the acrobat on the resonating wooden floor add to the '*virtual reality*' making the spectators feel that they are unwittingly caught in the middle of a contentious duel. They gasp and scream in wonderment, or fear; evades, ducks, and squats at the moves; Mothers grab their children and scamper out of the dome. Kindly, the troupe cuts short their performance as the audiovisual effects become too hard for many to bear. In any case, it has been a spectacular show of finesse, artistry, agility, and aerobatics.

The show at the courtyard is not over yet. It gives way to the Arabian sword dance '*Ardah*' by men. The dancers wearing traditional dresses and leather ammunition cross-belts stand in two rows, shoulder to shoulder; begin dancing to the rhythms of beating drums. This has been their traditional national dance they used to perform in preparation for the imminent encounter with the enemies on horse backs in the vast desert battlefields in their war history. It shows the valor and heroic spirit of those times. While singing, they sway back and forth; side to side; shake their shining sharp swords finely in constant up and down motion without hurting their comrades or bystanders. Right away, [as the say about the 'Liechtensteiner Polka':"*Da, bleibt doch kein Liechtensteiner auf seinem Platz...*". No one in Lichtenstein stays still as the polka music plays, and dancing begins!] the Sheikh and his sons jumps up and join in. They choose the glittering swords, form a row of their own; gracefully bent the knee and bent forward; lift the sword as they straighten up; then lower it in a rhythmic repetition while the drums beat in gripping crescendo. The wives get excited; take advantage of the rare opportunity to cheer their men and show their unbridled exhilaration. They clap in cheer; clasp in suspense; grip the hands of the '*sister wives*' in solidarity, and grasp them in great gaiety. In culmination, the Arab musicians play music by the famous composer *Aram Khachaturian* from his ballet '*Gayane*'. The fast rhythm, representative of the galloping of horses is played on the '*dombeck & Rique*; the hooves stomping the thick red sand is played by the *Oud* (5 pairs of strings tuned in unison in the 4^{ths}: C, F, A, D, G, C, and a single bass string at the end): the clang of the swords by the combination of *Rique* (tambourine). *Qunan* (plucked zither) provides the ambiance of birds flying away in fear. It's an extremely rare repertoire of music to be played with the Arabian dance. Usually, they only sing poetries or folks themes. Times are changing; there is no '*muthawa*' here to enforce the edict! It shows that life is better lived in their absence!

The VIP guests are escorted through a floral entryway laid with carpets that are intricately woven with historic Mughal & ancient Indian arts, symbols, and hallmarks. The archway vestibule connects the two realms (sectors); the exterior and the '*foyer*' (lobby). Up on arrival at the vestibule, the special guests are ushered into a '*holding room*' to stretch out and relax from the long stretch of the parade. They are not at all used to walking other than going around posh shopping malls avoiding the stare of stern looking, frowning Muthawas. Oh, this is Marathon for them! They are now presented with Indian spray perfumes extracted from plant, nature, and flowers; offered refreshment drinks. They catch their breaths after the longest walk ever undertaken. Meanwhile, inside the hall, the audience are comfortably seated and patiently waiting. With the plenty of novel things around, they peruse and ponder the surroundings; fiddle with the gadgets and controls they find; learn about the new technologies put in use there; expectantly, wait for the grand concert to begin.

At last, as all the members of the Sheikh's family make their expansive entrance to the hall, the crowd cheers enthusiastically. They do the hand clapping in the classical Karnatic '*Thala*' (Rhythm beats) of the '*Adi Thala*'; sounds like **Bhum** [P·P·]Pa-**Ppa**. Adi thala is the primordial (earliest/original) and the most popular thala in Karnatic music. This 8 beats cycle is roughly similar to the 4/4 March rhythm in Western music (Indian music is prolific in ragas & thalas. There are 72 ragas and 175 thalas in Karnatic music. The rhythm beats appear in groups of 2,3,4,....with strong and weak accents among them). The patrons acknowledge the great reception and enthusiastically wave to the audience, periodically switching to the traditional Indian "Namasthe' greeting hand gesture. The guests suddenly become aware what they are really missing in their home country. They appreciate the spontaneity and natural behavior people habitually make. The Sheikh's family take up seats in the entire levels of the balcony seats reserved for them!

Everything so far has gone smooth and as planned, though late on time. That's not unusual in this part of the world. Time has to wait for the convenience of the people and their deeds. There's no so much fuss unless you must catch a flight or train. At the moment, nobody seems to mind the delay at all. The ambiance is fabulous and invigorating! What is going to happen soon must be as exhilarating as well. The general mood is electric and anticipative. Time gives people an additional moment to sink-in what they are experiencing, relish the moment and the next.

Without any further ado, the front curtain slowly rises. The concert band is seen fully seated in their respective positions ready to play music. Everyone is looking towards the backstage entrance. A properly attired *bugler* stand upright, head high, shoulders back in front of the back-curtains with a bugle in his *gloved* right hand. The bugle is a wind instrument looking like a *cornet* without any valve; sounded by the vibration of the lips against a cup mouthpiece that can play only certain *note intervals*. The audience wonder what he is up to. Some rustles arise. Unmindful of it, the bugler steadies himself and blows a call to attention of an announcement of an impending event by playing a brief interlude from the recognizable tune by George Frideric Handel, "*The arrival of queen Sheba*". At the end of the ceremonious bugle call, the back-curtains are being drawn to each side by two ceremoniously costumed ushers in typical colorful Indian Turban (as the "greeters" at luxury hotel entrances). Elise is seen standing in the middle. After the curtains are fully drawn, she makes her way to the stage in grand style, pomp, and ceremonies, to the great delight and standing ovation of the audience. Gracefully, she strides to the podium in front of a golden microphone accompanied by the euphoric interludes and sound effects played by the band. The Sheikh likes the color of gold. It signifies the beginning (morning sun), brightness, hope, and certainly, the opulence he is accustomed to!

Angeles Mathappan, the '*Emcee*' (*Master of Ceremony*) greet all the guests, VIPs, celebrities, and the audience in the most fitting ways; cracks a few new jokes. Without further ado, soon, he yields the stage to Elise to start the much anticipated concert.

The band plays the instrumental prelude (A short introductory instrumental piece to the main piece) with great dexterity and warmth. Elise is now ready to begin her song. She always maintains a familiar routine, that she tells the story of the song in a mellow voice in a poetic recital. Every time when she does, her genuine emotions break out in full display. The orchestra continuously plays the background music to her dramatic recital with fitting sound effects providing the true ambience, just the same way as we are used to hearing in classic movies.

The audience reacts to it in great compassion and empathetic reactions. Its magical effect is that she has already won over the hearts and minds of the audience, even before starting her song. "Well begun is half done!" seems to be the chosen maxim.

The prelude begins with a delectable introductory violin piece, handing over to the piano for completion; returning to the violin who presents the '*tonic note*' (note upon which all other notes of a music's scale is based) to the vocalist after a delightful improvisational arrangement. The flute and Tabla joins in with the saxophone; the Veena is plucked on the treble string; it vibrates and resonates its high-pitched dulcet tone which sustains for a lengthy while before gradually falling silent.

Finally now, it's Elise's turn. She starts with her first word "Say...!" sung in a soft, long-drawn-out full note in subtle and delicate '*ornamentations*', (gamaka) which in progressively measured steps, fades. But sustains and floats much like the fragrance in the air; not wanting to fade away fully; making listeners look around which way it is going away. To describe the effect in a metaphor using familiar scene from life: "like children blowing soap bubbles in the air that spread and drift; float and linger around; pops and burst softly one by one, disappearing in a vacuum of thin air around them".
Only Elise is known to be able to perform this enigmatic voice effect! For many savvy music enthusiasts, this is actually more absorbing and captivating than the '*overture*' she recited from her manuscript!

Several sound effect engineers tried to simulate the sound using several soundboard effect manipulations including the popular guitar effect "*Wah-Wah*". But everyone failed; ultimately attributing the inimitable effect to Elise, admiringly calling it the "*Elise's float*"!

It is partly accomplished by an audio technique, designed exclusively for her by her cousin Tomichan who is a Broadcast engineer who excel himself in sound manipulation techniques. He used a special microphone array of omnidirectional microphone, one in the middle, and 3 'super cardioid' directional microphones at the left, middle, and right sides. It looked like a '*mouth organ*'. (Cardioid & super cardioid refers to the shapes of the sensitivity pattern of special directional microphones with respect to the location of sound sources. Their acute directionality rejects sound away from its front and axis, such as band instruments, room reflections, ambient noise, etc.) They carried out experiments with different models; used several trial & error methods and perfected this effect. To produce this effect, Elise drag her voice continuously and delicately from side to side. She varies her relative position from the microphone array and moving smoothly across from left to right. When Elise makes her patented '*float*', people think she is simply blowing into a peculiar harmonica from one side to the other! It feels like the sound originated on the left side (the left loudspeakers sound louder; then feeds into the overhead, which finally transfers it to the right side speakers which become the only active one. Moments later, they also fade slowly to complete silence, completing the effect.) The sound is perceived as exited out of the hall through the right-hand exit door!

The directivity of the microphones can be either exponential or parabolic. [The following explanation is made for the sake of mathematicians and because of the fact that music and sound wave propagation usually follow mathematical equations!
The sensitivity of parabolic curves is based on 'quadratic functions $y= a x^2 + b x + c$ ' in which the increments in 'X' cause proportional changes in 'Y' ; the curve looks symmetrical, and smooth. Whereas in exponential curves: $y = a * x^b$. The difference in 'X' values cause steeper and steeper changes in 'Y'. In other words, when the sound source moves away from its dead center, it decays much faster-exponentially!] She had done several voice exercises and evaluated the response. She will have this session recorded and will listen to it in the hall sitting at various seats while her recording is played in continuous loop. Any imperfection will be duly corrected and improved. For such important reasons, Elise always carries this custom rigged microphone array to all her concert performances. It contains an electronic chip using the algorithm described previously and includes a number of personalized voice effects as well. She insists that it be used instead of any readily available microphones, solely because she wants to keep control of using the extra effects as how, when, and at what time. Not all the crew is amused with it; but they comply, as the unique quality and execution of it reflect on their skills as well.

The audience react to this voice effect accordingly with unconcealed awe. It's unfathomable how just a single note can capture so much attention and praise.

As usual, she began: "*Say...! just say....! oh, oh-oh,...oh...!...just say...! that I heard you...!*" in a delicate, captivating, appealing, melodious voice, articulated in corresponding body language and facial expressions, only an authentic human feeling could bring about. It is her distinctly characteristic style. That's what the '*viewing audience*' came here to see and hear as to how she performs this delicate, musical magic. They say "*seeing is believing*!" It's simply magnificent, and mellifluous! A few recording artists and musicians had argued that it was done by means of electronic aids of voice manipulation, synthesizer, or simulated sound. Some audio labs did become intrigued by the *voice phenom* that they conducted several tests and examinations. An expert audiologist recorded her voice in the lab and certified that the voice is naturally produced. There were several audiograms, oscillograms, and spectrographs to prove it. Typically, the audiogram tests the hearing ability of both ears from the lowest pitch of 125 Hz to the highest pitch 8000 Hz. Elise's result exceed and extend beyond the average test model for pre-teen children. The 'speech *banana,*' the so shaped area of graph where all the normal conversational sounds are heard is found to be extraordinarily sensitive. There is strong relationship between hearing and speaking, meaning one has to hear well to speak well! Indirectly, the speech banana response diagram also signifies a person's ability to learn languages. The sounds *M, U, O* occur in the rather low frequency range below 600 Hz; ***EE, P, SH*** in middle frequency 1000-2000 Hz; **TH**, **F**, **S** in the high frequency of > 5000 Hz.

The visual representations of *oscillogram* represents waveform of sounds produced shown as a color graph plotted **Amplitude Vs. Time** in sec; whereas, the spectrogram is a color graph of *Frequency Vs. time*, showing the energy in sound frequencies in varying intensities of color. A Japanese consortium representing the manufactures' Association of digital music vendors bought the rights of this '*sound effect*'. Digital audio converts audio into numerical values after taking samples at specific *sample rates*. They copied the wave form; applied the mathematical concept "*Fourier Transform*" to filter /split the signal into its constituent frequency components, and later digitized the waveforms. Once in digital form, it can be distributed for public use as the signal can be seamlessly decoded by '*codecs*', a hardware and /or software application that can encode or decode digital data streams, automatically detecting the involved algorithm.

Chapter: 21

A mellifluous song

The eventful concert

Say,... Just say...!

Say,.. Just say...!

Oh, please,... Just say!

That I heard you

When you hear my song...

Yes, I heard you...!

A song, I sing for you

For the life you've saved

The gift you 've given!

This song I'm singing...

Is for you!

To hear and say...

For the life...

You once saved

And for the good deed...

You've done

Long, long, ago...

Just say..!

I sing,

As loud...as I can

As sweet... as I can

To reach you

Wherever you may be...

Or how far away you're!...

To hear you say....

“Yes, I heard you!...”

My invisible friend...

The long lost friend..

Oh, my lifelong friend!

Say!

Just say!

The simple words

I long to hear

say!...

Just say!....

"I heard you!..."

When she hits the highest passage, the green colored fluctuating audio signal waveform of the same is superimposed close to her mouth as if she is blowing out the complex sound waves along with her voice. The annotation on the Television monitor marks out the dB amplitude level of the sound and the name of the note which is #F5. It is 3 whole note lengths in duration of 4 counts each. The image freezes on the monitor; highlights and expands the waveform into full screen zoom, showing a big and clear view of the waveform envelope with its entire subharmonics, when Elise produced the enigmatic vibrato effects. The viewers speak words of amazement of '*awe, vow, disbelief, dismay, wonderment, and admiration*'.
At this instance, the crowd becomes overly excited; they whistle the "*Kurava*" (ululations- produce howling sound effects of joy or exultation by alternating rapidly between two high-pitched soprano sounds- commonly seen in Asian & Middle eastern cultures)

After each '*refrain*' (A much repeated part/stanza/chorus of a song) there is an interlude (a short intervening tune connecting stanzas) of a series of instrument solos, in which the instrumentalists play their renditions and variation of the same musical theme, or phrase. They try to win over the audience and outshine each other by introducing a variety of articulations, and ornamentations. Elise continues her mastery of voice to the end of the song in glorious style and melodious tune faithfully to the last passage.

As she is ending the memorable song with its motif (a brief musical piece containing thematic 'catch word' or structure) "*Just say I heard you!*", a musical voice came from the back of the hall, as if it were her own echo in corresponding '*tenor*' (the highest natural adult male voice) voice:
"*Yes,... I heard you!...*".... "*my grie..ving..., Sis!...*"
Initially, the audience assumed that it's a clever sound effect, as they have heard plenty already in this song.

To great amusement of everyone, a person is seen standing on the isle under the lower balcony and in front of a '*floor-standing*' microphone which is flashing a faint '*yellow*' indicating that it is a '*voice activated*' device! As in a duet, the gentleman sang back in his tenor voice in a responsive musical tune, as in '*antiphony*' (a responsive, reciprocal, interchange; conversational exchange) of sorts to Elise's endearing emotional plea. The sound was coming only from the rear left and right speakers unmistakably identifying its location and the source of the sound. Suddenly, everyone stood up and stared towards the rear, like a herd of deer with its ears erect and turned to the location of an approaching danger. A light fades up and cast on the person, probably done by a vigilant and thoughtful specialist in the lighting crew.

Automatically, a camera pans and focuses; smoothly zooms in on him showing his full figure, his hand still in an extended greeting gesture in the direction of the stage towards Elise, something like Saddam Hussein was famously known for, while greeting the cheering citizens. Elise is stunned! Surprised, and stupefied!

The orchestra stopped; no sound came out, as if the electric power to the sound system has gone off. There is absolutely no sound, or movement in the hall. The constantly resonating dome did not fail to reflect the enigmatic "*sound of silence*'. It was the most stunning sound effect the audience heard today among the very many!

Nothing like this has ever happened in the history of a concert. A silence that's powerful and captivating! It blended with the music so well; a long musical "*pause*', a lull of suspense, wonder, dismay, and a suppressed emotion of curious kind, simultaneously occurring in one poignant moment! The 20,000 listeners hold their breath at the same instance. The air is thick, as nothing is drawn in. Also, there's no air movement in the whole hall!

Initially, Elise thought it was some prankster (a practical joker) drawing attention to himself or perhaps merely an overzealous fan.

Immediately, a voice prompt came onto her concealed *intercom earpiece*, tucked under her groomed, springily, thick, and dark hair, similar to that of newscaster on the Television broadcast desk. It's a one-way communication link; purposefully designed so that she can only hear and cannot talk back; there is no microphone attached to it, to speak into. It's actually meant for private channel conversation and thus may not be aired inadvertently! She cannot talk back; it is designed that way, and in '*one way*' only!

"The Producer speaking! Ask the gentleman if he would care to come on to the stage!"
Elise obliged tacitly; faithfully followed the given instructions as naturally as if it were her own deeds, acting on her own volition and reflexes.
Confidently, the gentleman, strolled towards the stage; climbed the steps in a jaunty, timed, sporty hops, a style remindful of former US President Obama, as he climbs up the boarding ladder to '*Air Force One*'. He was guided to step closer to Elise facing her. He was formally dressed in western style: in pants, shirts neatly tucked in, wearing a sports/ travel coat with a tie in classical knot in the latest Broadway fashion. He has smartly combed up and partitioned curly hair, sporting a carefully trim-cut beard, the so-called "*Arabian Sheikh's beard*", named after the long-served oil minister of a gulf Kingdom.

She thanked him for readily accepting her invitation in a gracious manner.
The voice prompt from the producer came on, once again.
"Ask him if he is alone or with anyone else; play along and follow your instincts! Is it clear, Elise?"
Elise showed no nervousness or panic; she felt sanguine! (cheerfully optimistic); There is somebody to guide her through. So, she played along, not knowing what it may lead to.
The same time, she is free to follow her own script for the first time. Fine, from now on, it's literally her own show!

He answered briskly: "my immediate family-my wife, son, and daughter are with me; we were sitting together... enjoying your lovely voice in this awesome, prodigious, beautifully sounding hall... until you broke us up!" uttered in good humor and with an amiable smile. The audience also felt amused and clapped showing their strong approval. She hesitated a little bit; felt some familiarity, faint resemblances, and indications, true to her distant nagging memories. Quickly, she dismissed all of it as a bizarre '*wishful thinking*'; pretended to be just normal.

She paused, a little perplexed as to what to say next.
Then asked spontaneously if his wife would also do what he did.
He replied politely: "Well, I guess... the best way..."; then hastened to say "You could find out by simply asking her directly!"
He replied in non-committal way.

Elise dared and spoke right into her hand-held Mic and asked:
"Dear Mrs. would you please care to accompany your husband on the stage?"
The Mrs. smartly stepped up to the same *mic* her husband used minutes before; responded: "Yes, of course!,... gladly, indeed ma'am...!"

The isle lights lit up and lead her to the stage.
A pretty lady with thick propped up hair bundle, adorned with marigold and yellow flowers placed in concentric circles, as seen on the rear lights of the latest model of Korean cars, and white jasmine flower bunches neatly filling the '*valley*' of her partitioned hair; walked up looking confident and smiling, in modest measured strides, towards the stage.

All eyes followed her along, all the way. "Perhaps she is coming from a family wedding reception" a lady sitting on the aisle seat murmured with a hint of envy. As she reached the top of the stairs, a '*run /deck*' attendant placed a peculiarly shaped gray leather case on a hastily brought out '*table-top*'. Gently, he pointed to it and waved her attention towards it, like a courteous attendant would do.

She recognized the distinct "*Hohner*" insignia (a symbol / trademark) on it and touched it. None in the audience, nor on the stage, or for that matter, even Elise could comprehend what it was all about. Surely, none of it, is her own making! Perhaps, a scene from the *game show* on American TV.

Once again, the producer spoke in to her earphone "Be calm; ask Mrs. what is in the case; is it a magician's box?" Again, Elise followed the instruction '*verbatim*' (using exactly the same words).
"Are you a magician?.. Is it a magic box?"

"No,... Oh, No...I'm not!...Not sure what resides in it; hopefully, not a beast, Jumping jack, or Cookie Monster.

Looking around, and keeping a distance, she said: "Ok,...I'll be the '*scape goat*'...You want me to find out?...Let me open it!... I can do it...boldly of course,... Let's find out! ... 'Hope it's not a skunk,...watch out, now!" Please give me a little room to duck away,...in case...!

Masterfully, the Mrs. opens the box as opening a leather trunk or '*portmanteau*' (a leather case made for packing travel necessities). They have identical buckle-type fastening mechanisms; one on each side; slide on the keyhole buttons outwards, the metal latches spring out. She seems to have done it many times before; swung it open into two sections which are hinged together at the back. She was surprised noticing the address label and the insurance identification pasted on the inside back-flap of the case in her husband's own handwriting. In great amazement and excitement, she shouted: "It's my husband's accordion!"...looking towards her husband in disbelief, and inquiringly.

Her husband is also seemed amazed and had an incredulous stare on his face. He knew that he had carefully packed it himself in a cardboard box, obsessively stuffed it with a lot of packing bubbles and Styrofoam corner blocks. The instrument is to be handled delicately while transporting. He had carefully, taped down all the fragile chord buttons so that it won't collapse and get damaged in transport. (an important step in accordion packing & shipping!). It was appropriately labeled "*fragile, handle with care!*" He just assumed it was a case of "*lost & found*". It could have been retrieved just because of the shipping label on the box. "Thank God, it's an expensive '*legacy*' musical instrument. The same is no longer made!" he mused.

Their children stood up from their seats and leaned over for a better look. The cameras captured their consternation in full measure. Now, everyone is puzzled. No one seems to know anything at all! Everyone wanted to see what is really in it; what it looks like. Nothing jumped out, or flew away; heard no murmur; it's not pigeons nor a mouse; it's not anything to do with magic. It must be an inanimate object, for sure!
Yet, this is worse than a magic puzzle!

Elise asked "Mrs. you play this instrument?"

"No, my husband does! In fact, he plays your tune quite often;....although he calls it by some other Spanish name. I never understood the correlation. I've asked why he liked it so much. He said he can relate to it well, and play it well too!"
It spiked Elise's interest and optimism of a potentially interesting serendipity (An unsought, unintended, and/or unexpected, but fortunate, discovery).

The garrulous Mrs. continued: "I told my children that Elise was singing their daddy's song. They said that they did already know that much with an air of sarcasm!" (a mocking remark). "Sometimes, they play the same song together, each playing a different instrument.
It's a very nice song ;.... it's quite moving ,.... particularly sweet in your voice!" She continued rather excitedly.

"Can you kindly tell us who plays what?" Elise asked in some haste.
"As I already mentioned, my husband the accordion; my son the big Piano; my daughter the fabulous Violin."

"And you?"

She blushed. “Well, me?...Oh, well,...there should be someone to listen to it all; ...the noise I mean, you know? Faithfully, I fulfill that duty. Sometimes, they ask me to turn the music pages over. I end up turning the wrong one”.
“Mummy, you turned the wrong one!” my daughter screams at me annoyingly.
“I escape with a sly smile;... no damage done, except for my bruised ego!” She plays on, trusting her musical memory from then onwards. Well, she learned violin by the “Suzuki” method mainly, but not solely by listening. She reads music sheet as well”.

“You mentioned they are all here; would they also come on stage, following the good examples the parents have such gracefully set?”

Without waiting for an answer, Elise announced:
“Would you please both come to the stage and join your parents?
Hello..., where are you, siblings?... stand up please!...will you?
Oh, there you're!... come on... come right on to the stage!
Let's make it a family reunion,... of sort!”

“If I break something up, I should at least try to put it back together!” She said it slyly, and poignantly! The members of the audience chuckled. They just remembered what the gentleman remarked earlier; possibly a *repartee* to him; in his own words.

A handsome boy approached from the left, the Piano side, and a cute, pretty young girl approached from the violin side.

The audience begin to feel that they were witnessing something new and special in Showbiz and concert. They are used to only listening to the music, and go home. That was it; just some music; no drama; no intrigue; no hoc-us-pocus; all straight forward to bare bone; nothing to ‘*ruminate*’ up on (to go over) afterwards. Most performances had been somewhat stale, without any other distractions, deviations, persuasions, or excitement. By far, this is definitely different! And it has only just begun!Everything is quite different in here, from A-Z!

The audience broke into a rhythmic applause encouraging the youngsters, as the siblings approach the stage in synchronous steps and turning towards the audience, smiling in recognition, nodding head in acknowledging their appreciative welcome. The young lady's bangles jingle a musical clang, something like a '*bell shake*' as usually heard at church services, waking up the tired devotees who had unwittingly and momentarily fallen asleep, as she waves to the audience. The audience is amused by the spontaneous happenings and the interesting dramatic stream of events it's producing.

Elise stepped forward and warmly greeted each one of them; straight away asked if they would all play together,... anything at all- from nursery rhymes to anything you like...; Come on,...don't be shy! She said encouragingly. The violinist in the orchestra promptly handed his violin to the Miss. She took the violin in her left hand, the bow in the right. While doing so, instinctively strummed the strings in quick succession (pizzicato-plucking/pinching style) as if it were a guitar, followed by a couple of "*trills*" a "*slur*" with one or two down bow on either side of the bridge, to end with a few delectable "*spiccato*" (bouncing of the bow after each note), all executed in exquisite style and with consummate ease. She was checking the tuning of each of the strings E, A, D, G. (E, the thinnest, produce the highest pitch, the others proportionally in lowering pitches in the increments of the 5^{ths}.) which every musician usually and habitually does.

The audience becomes eager and excited; they anticipate something interesting is going to happen. They acknowledged her skill by a series of whistles. She bowed to them in polite acceptance, and polite, interactive gestures. Similarly, on the other side, the Pianist got up from his stool and escorted the young boy to his piano. In the act of sitting down, he played a '*glissando*' followed by an "*arpeggio*" (playing individual notes of a chord up or down the keyboard in sequence) of C minor starting from the first piano key F_1 to the last key C_8 in quick succession in ascending and descending order. For its elegant execution, he casually applies '*hand over hand*' method as the drivers negotiate successively quick left & right winding turns. The boy accomplished th act in fluid horizontal upward and downward motions. Beautifully executed!

Both seem to have completed their '*warm-up*' exercise; show familiarity and comfort with the instruments and the quick events they are now involved with. Or, the boy was perhaps trying to emulate his younger sibling, wishing to earn some accolades for himself. His '*show of skill*' was equally phenomenal; also recognized with applause!

Elise asked "Are you guys now ready to play something for us?" The mother, frantically pointing to her watch, gestured to them to hurry themselves up.

The father selected the *musette* (3 reads beating together in producing a phenomenal 'bleating' sound) voice register (the closest to Elise's voice) on his accordion, played the first measure of the song as they usually do it together "*Say...!......*" in a soft appealing tone; the tonic note gradually and dramatically '*fading up*' and then likewise '*fading out*' smoothly in reverse. They played it in the style of "*Elise's float*' which they consider as another innovative embellishment of "*overtone singing*" (a.k.a. harmonic /polyphonic overtone singing).
[*It's a difficult and complex 'voice technique' attributed to nomadic Mongolian region and tribes. It's an art of singing two notes at the same time; a voice technique whereby the resonances produced in the throat and mouth are combined with the manipulation of tongue, lip, and jaw movements in such a way that individual* **overtones** (the frequency harmonics that occur above or multiples of the fundamental frequency) *become so loud or rise & fall that they are perceived as individual tones.*]

In the rendition played by the trio, they played the *motif* in '*counterpoint*' (meaning the simultaneous sounding of two or melodies, represented on different staves on a music sheet). The violin played '*voice-1*'; the accordion in its musette register played '*voice-2*'; piano added a number of *tuplets* (*a set of short notes subdivided within a beat*) embellishing it in soft and subtle tones. Elise and her orchestra are intrigued at the technique and the ease with which they are renditioned. Immediately, the bandmaster recognized its beauty, and made a mental note of it!

The violinist followed the next measure, "*just say!...*" after a dramatic quarter note-length '*full rest*' (interval of silence in music) in a sweet, mellow, enigmatic tone adding a delicate finger '*vibrato*' at the end. The Pianist complemented them by playing both the measures "*Say...!......; just say!...*"

likewise with his own rendition, by engaging the '*Sostenuto*' (Middle Pedal) to sustain the ending note a measure longer.

Now, it's clear to everyone what the trio is going to play. Elise blushed and at the same time elated. She is also overwhelmed with emotions, surprise, and joy. She is astonished at the passion and liveliness with which her iconic song is played by the guest musicians. She never thought that her music has a wider appeal than what she had imagined. Concurrently, she is in a dream world of her own believing more and more that her song may have finally reached its intended destination, and heard by the person, she has been singing to, all this time and that person may be already here now. She felt that now the '*tables have been turned*' on her; the song sounds like a '*serenade*' (*a song performed to honor or express affection and love for someone special*) in her honor. Tears filled her eyes; the drops involuntarily trickled down her cheeks. The camera is quick to capture the emotionally charged moment, the natural behavioral component of emotion (emotions have 3 main components: physiological, neurological, and cognitive) in its authentic expression, and intensity. The camera shows the size of the drop in extreme close-up and with magnification and dramatization.

The Mrs. was also greatly moved by Elise's emotions; handed her a fresh facial tissue and bloated the tear drops from her face in a loving act of kindness; held her hand for a brief time until Elise gathered her composure. She whispered into Elise's ear that she may join them at an opportune time whenever she would feel ready for it. Elise agreed with a consensual nod, and a sigh. In the meantime, the trio played variations of the initial few measures and kept the music going, building up the mood and maintaining the tempo. When one plays the lyrical part, the other two play the interludes. They alternate their roles for the succeeding verses /stanzas. The left hand of the pianist continuously play the beats & rhythm as usually the tabla does. The right hand complements the melody the other two are playing, adding occasional '*fill-ins*'. The listeners feel that the melody and the rhythm moves in synchronous steps and leaps as a horse's gallops and the carts' clanking of the wheels. To some it felt like the high-kicking military parade marching in familiar formations. They distinctly feel the movement. The trotting (trot:run without galloping) of the horse makes a sombre, steady rhythm! At certain musical moment, all join together in a chorus playing the '*motif*' in a distinct instrumental '*declamation*' (an act of rhetorical delivery) of polyphonic sounds. Each time, they try out another variation in the overtone singing style. This time, the accordion plays '*voice-1*' as a long held out note with his little finger, and half of '*voice-2*' with the rest of the fingers in softly blending legato. The violinist adds the rest of the voice-2 notes complementing her father's musical tones. They communicate with each other in subtle musical terms and intuitive ways. The listeners are enthralled! Immediately, they recognize an extra dimension of the same music.

Elise mustered up some courage, looks towards her husband, the orchestra's violinist and then to her brother who is the band's pianist in the orchestra. Elise steadies her composure; receives a reassuring signal. She pauses for her breathing to normalize and have a full breath; gestures to the instrumentalists that she is ready to join in. Elise simply starts on the key, the Miss. violinist has presented to her after completing her solo '*intro*' piece. Amazingly, that was her own high-pitched key of $D5^{\#}$! *(D sharp high in the top octave on the keyboard ; R3 in Karnatic music).* The Miss. violinist had correctly identified and memorized the '*tonic*' note (beginning keynote of the scale) when she heard Elise's opening song! That's an exceptional skill, indeed! Elise is surprised; smiled to the violinist in appreciation as it spared her a cumbersome '*transposing*' of the musical scale to match with the guest orchestra. On hearing the smooth transition, the audience stood up in hushed silence and admiration. They felt that they are experiencing a musical '*transcendence*' (going beyond usual limits), and spontaneity; something in the realm of the mystical world of "*Gandharva Sangeetham*", the divine music gloriously praised and fervently followed in ancient Veda.

Now, Elise feels really at ease; picks up her sweet solo voice in usual style and completes the first stanza. Usually, it is followed by an orchestral accompaniment. Recognizing the fact, Miss. violinist attempts to return the violin to the 1st violinist of the band. But already, he has his 2nd violin in hand which he always keeps by the side, since he once unceremoniously broke a string while playing a solo, some time ago in another concert. He completes the fill-ins together with other members of the original band. The musicians speak the same '*lingo*'! (the specialized vocabulary of a particular field or social group). They finish the part with a tabla improvisation in "*Adi thala*" (the most used Beat rhythm in Karnatic music) before handing over to the trio who repeats the melodic part, and in turn, present it to Elise, for the concluding part "*I heard you!....* Elise is pleased to hear another delightful variation of her own song. Reinvigorated by their rendition, Elise sets herself up to sing out the concluding final lines in a *declamatory* finale. With her steady and awesome *coloratura soprano* voice, she sings "*My... in-vi-si-ble **friend**... The long...lost **friend**!...My life...long **friend**!* in one exceptionally long, ***held-out*** sequence, and breath. In the process, the critical passage '*My life...long friend!*' hits the highest note with sensationalistic crescendo, and a tormenting, nostalgic undertone. The instrumentalists struggled with ways to accompany her. *They fell by the wayside*; smartly, let her carry on alone in her incomparable, pulsating solo voice that itself created an enigmatic musical stress, and intriguing conclusion of the song.

The listeners are flabbergasted! Ecstatic rounds of clapping and jubilation spontaneously ensue; thunderously circulate around the hall. Elise shares the excitement and euphoria of the moment with all the rest of the musicians in great gratitude, humility, and magnanimity.

At this instant, streams of compliments & comments started to pour in via the social media platforms. The audience read them out loud one by one as they appeared in real time. In exhilarated descriptions, the comments describe what they have heard, seen, and experienced.

The organizers, Television, and online video streaming viewers are also delighted with the music and the unscheduled, adjunct guest appearance of the family trio. They display their unbridled elation. ***The opening concert billed as the Grand opening gala' surpassed all expectations!***

This concert is '*simulcast*' and '*live*' on satellite Television, viewed by the Malayalee diaspora spread across continents, countries, and islands around the world. It's also being '*live-streamed*' on Internet & YouTube. Some sailors can receive the program on their merchant ships and cruise ships. A number of very thin, mildly curved Q-OLED (Quantum organic LED) screens are embedded on the walls of the Amphitheater, seamlessly fitting with its own aesthetic curves. It provides in-obscured Ultra High definition large screen display showing the actual on-air transmission beamed out of the Broadcast center. The screens provide only the video, devoid of sound, obviously to avoid echo, audio delay, and other annoying artifacts. It was shown with the intention to enable the audience to see the studio generated titles, crawls, roll inserts at strategic '*points in time*' of the concerts along with an array of eye-catching digital visual effects, such as dissolves, wipes, spin, tumble, fall, rise, twist, turn, zoom, masks, shadows, reflection, reveal, keys etc. for the Character /Title Generator titles, as well as scene switching effected by the *Video Production Switcher*.

The people on the stage are also given this facility on the floor-mounted screens on both sides of them to get visual feedback. This enhances their stage awareness; assists them intuitively modifying individual performance according to the current responses they are receiving. Improvisation, spontaneity, and audience stimulation are typical in any artistic performance. At this very moment, the social media continues to show spikes in activity.

People had been writing comments; ratings, reviews, reactions etc. They are frequently displayed on the screens. Seeing the surge in participation and positive ratings (thumbs-up signs) as well as viewer responses, the Television executives smartly announce pre-empting (take precedence over other) the immediately following program so that this concert can run its full natural course.
The viewers praise the decision!

The song "*Just say, I heard you!...*" ends in a dramatic, emotionally charged, captivating fashion and rendition with hitherto unheard of musical excellence and added significance.

The performers receive standing ovation and praise. Elise gives the violinist an appreciative hug, directs the applause towards the pianist and the accordionist. There is no eye in the concert hall that is dry!

It's not just a concert, but a meeting of the minds; participation and collaboration by unknown and unfamiliar musicians complementing one another in uncanny, musically spontaneous ways. Just magnificent!

It is emotionally charging in a way that only the true Indians seem to be capable of feeling. How they pooled together their individual talents and pulled out the best effect in a music that is so familiar to everyone. It is simply amazing. It's not just music but also the bringing together of the goodness, tenderness, and humanity within people. The interactions, mutual respect, and admiration shown between the performers and the listeners stand out. There were many wonderful episodes and enjoyable aspects in the extempore musical performance to behold and cherish.

This is a unique event, perhaps with both musical and sentimental overtones to it. The audience is relishing it; tremendously! They enjoy the excitement, and magic of the moment. A musical passion has been set ablaze. The audience want more!...

Everyone is caught up in the overpowering spirit of the moment. Indeed, they clamor for more; the same or another song. Some missed recording of it on their smart devices. Meanwhile, Elise received the official go ahead. She asks the family musicians if they could do just another one. The accordionist ask if he can have a wireless Mic, preferably a head-worn one. Promptly, he is fitted with one. It looks like the request is heard loud and clear. With that hope, people calm down and wait.

“Well, we shall play a wonderful piano composition by Ludwig Van Beethoven. As a secret admirer and lover, he composed this music for one of his beautiful music students Therese (Elise, is a nickname to disguise true identity) of a noble, aristocratic pedigree, and status. The music was simple enough but strong in emotions. Beethoven hoped that she would understand and respond to his true and affectionate feelings encapsulated (as contained in a capsule) in the composition. Though he gave it to her secretly and quite discretely, it was later discovered in Therese’s music collections and released to the public. No one knows if Beethoven lived long enough to know about the *revelation*. It would have been an unbearable *double blow* to a proud maestro!
Well, this so-professed love was the composer's *muse* (a source of inspiration) and theme of the song!

It's a very beautiful song; so much so, I wanted other people who don't necessarily possess any knowledge of Western classical music also be able to sing and enjoy the same. I wrote English lyrics for it to reach a wider range of music lovers. From the outset, I knew quite well, it would be an arduous and challenging task to bring about the intimate and intricate feelings as expressed in the music, in scant words that I knew and the inadequate feelings that I had.
I was never in love; have been constantly searching for something I couldn’t get,... let alone, achieve!
Well, I was never satisfied with the lyrics I wrote; reaching my limit of ability, left it likewise as a “*work in progress*”.

Nevertheless, many found it quite enjoyable; you may take a look at it and determine for yourself!

Well, for now, I will play the melody, my daughter will accompany me on the beautiful violin, my son will provide the mystifying Bass rhythms. We hope everyone will enjoy listening to it, especially our acclaimed singer and the star performer of this majestic evening, the beautiful Elise with the great gift of a truly mesmerizing voice!....
We thank her and the audience for the cordial welcome accorded to us. We’re truly honored to be asked to play another tune for you!

It's titled in Deutsch "*Für Elise*", translated: "For Elise!"
"Yeh...? ...Heavens...Oh, my... good Lord!...
Elise exclaims; cheeks turn rosy; holds her breath wondering what type of music is it going to be; waits for its start in anxious anticipation!
Her mannerisms and reflexes are all spontaneous, dictated by inner feelings, not the way the producers wanted them to be presented.
It has its own natural, authentic beauty, warmth, and charm.
At this moment in time, she is no longer a performer;
Just a curious and avid listener; can behave accordingly and naturally!

..."By the way, my lyrics is titled '*So sweet as you are...yes you are!*'.
For now, I'll play the "*motif*" (a short rhythmic or melodic passage that is repeated), also the *refrain*, the verses that repeat periodically. I shall give you some words to hum or sing with, so that you can actively participate in. It's really simple; only one line! Let me just imagine that perhaps this is what Beethoven seemed to tell his lovely pupil *Therese*":
"*E-lise! hear me, hearme, hear me, hear E-lis!... Hear- E-lees!...hear me,-hear me, hear me,- hear me now!...* (sweetly and in retardando) *oh, my dar~ling...!*"
The accordionist sang a few times while his children played the *refrain-tune* in background accompaniment.

"I'll sing it one more time for our star singer Elise's sake who also shares her name with this famous music title. I hope she may lead you all in the frequent refrains... rest assuming she'll wholeheartedly agree!"

He looked towards Elise. She nodded in consent with an agreeable smile. Elise is wonder struck! The camera zoomed in on her face, uptight; her face blushed; her hand holding the Mic is seen trembling, also showing rising goosebumps.
She almost swooned! (To be overwhelmed by ecstatic joy)

The piano started, setting the mood and the rhythm in the characteristic slow, flowing, cautious manner, expressing love. The '*ostinato*', (a rhythmic phrase that is repeated throughout as a bass musical accompaniment) in *triplets* which sounds like: 'Pum pum pum-Pum Pum pum- pum Pum paa'; the violin joined-in for the next measure; the accordion completes the phrase playing "Pum pum paaa..." in '*ritardando*' (music: gradually slowing down) and a slow, melancholic '*dissolve*' adding beauty, resonance to show anguish and pain of a desperate, and '*broken heart*'. They repeated the refrain once again in the background while Elise lead the audience in humming the tune with the given words. Immediately, the alert *Character Generator* operator inserts the words of the refrain in a bottom TV *screen crawl* at each appropriate time. The audience happily '*sing-along*' using the words following Elise's lead and timing! A few of the loving hearts responded with a corresponding expression and feeling: a feeling of love and tenderness dreaming that their own sweetheart might someday ask for their heart in the same sweet, irresistible manner, and earnestness.

As a variation, occasionally, the pianist plays the melody in solo by softly and gingerly touching the piano keys to express the extremely tender and delicate feelings contained in the music. The melody and the accompaniment sweetly blend, raising heartbeats. The audience seemingly gasping for breath holding back tears. Some ladies are shown on the screen wiping off spontaneously flowing tear drops.

Each instrumentalist, serious and sombre, fully engrossed with their music, play their roles; the cameras switch back and forth. The musicians are shown in 2 way / 3 way / image splits; in *montage* (a composite effect assembling several elements of a scene) of 3D perspective, periodically adding yet another frame with the countenances of the audience's, and Elise's interchanging. The 3D *collages* provide an illusion that all the three people are facing and communicating directly with each other. The camera operators complement each other by showing the player's individual expressions, fingering styles, and the accordionist's '*bellow pull, squeeze,* & '*shakes*' like a big bird spreading its wings and displaying colorful *plumage* (the layers of feathers covering the body of a bird).

The instrumentalists swing and sway spontaneously; look sad and emotional. But certainly, they enjoy the music wholeheartedly, without which, they wouldn't be able to play as well as this! The young pianist played the solo of the chromatic notes spanning over two or three octaves in sudden but delicate succession, creating pulsating vibes in the listener's minds and body, so sweetly and tenderly expressing a lover's sincere and agonized pleadings.

Impulsively, it prompts some loving hearts in the audience slowly and slyly slide their hand towards their sweetheart's to express their own sentiment in a statement of love, and affection! Those who are grappling with rejection find in good company with the maestro. They understand the expressed emotions in the music so much more.

Abruptly, the music enters into an '*agitative*' stanza expressing waxing anxiety, distress, despair by playing tense discordant notes in a chaotic manner, feeling acute torment and anguish; then slowly transitioning into a serene consonance, and melodic passages depicting tenderness, love, sadness, affection, and hope he is capable of. All the contrasting emotions are wonderfully interwoven in dramatic and bewitching musical episodes with dissonance & consonance; harmony & disharmony; varying tempo- *retardando,* and *ritenuto;* crescendo & decrescendo, and *dynamics*. *Retardando* and *ritenuto* indicate the slowing of the tempo; the latter instructing to modify the tempo more suddenly, and extremely than the ritardando.

In the second repetitive part, the pianist lets the accordion play the rhythm pattern [*E, A, A ; E, A , C ; A, C, B*] in single tones on the *Bass* buttons (selected from the *2nd* row of ***Root*** button and the *1st* row of ***counter Bass*** buttons) with his left-hand, so that the audience feel the same sentiment in another set of '*rumbling*' (low notes) tones. The players coordinate well; go back and forth seamlessly, then blend in; deviate briefly; join again in harmony in delectable unison. The violinist helps to embellish the *sorrowful* part. The piano accentuates (to pronounce with stress) the *agitative* part, simultaneously playing the steady rhythmic structure of the individual notes in *A minor/E major* chords inter-changing, with his left-hand. The modulation of sounds when two or more instruments are played together, creates yet another listening sensation, literally moving the audio wavefront as the gentle breeze escorts the waves to the shore.

The sound of the music is just beautifully enchanting and delectably entrancing!

The song comes to an end, showing extreme pain, desperation, but accepting the inevitable outcome. The instruments join together in expressing the same sentiment in the best possible way their instruments and Beethoven's music can possibly express. That was brilliantly unique!

The listeners imagine that a wine goblet has slipped off from a bar table spilling all its contents onto the sparkling marble floor; the glass shattering into a thousand pieces. The broken parts strewn around all over the floor! The listeners feel the grief heavily on their hearts, but suppress it, empathizing with the *maestro* Beethoven in full earnestness. Collectively, they let out a loud sigh of genuine empathy, the theater sound system amplifies it in dramatic fashion! A sullen, sorrowful gloom, and sadness spreads in the auditorium at the end of the song.

Once again, the national psyche and philosophy reaffirm its long held belief: "everyone must invariably suffer some sort of peculiar or tragic situations in one's life!". Westerners may quip "*that's life, honey*!"

The bottom crawl of the screen shows posts & comments coming from, Kuwait, Manchester, Melbourne, Mombasa, Johannesburg, Abu Dhabi, Sharjah, Singapore, Kőln, Kuala Lumpur, Dammam, Dallas, Dubai, Dublin, New Delhi, New Jersey, New Hartford, Jeddah, Mumbai, München, Muscat, Lagos, Leeds, Toronto, Oakland,....Rome, Riyadh,...Jakarta... Bali, Bangalore, Berlin, Edinburgh, Cape Town, Achen, Amsterdam, Heidelberg, besides many local places in Kerala.

In their comments, some decided to share their tales in simple terms and language with the world, hoping to find a much-needed solace, resolution, or reconciliation with their own personal plights. A few claimed the music was therapeutic to find comfort in accepting their inevitable rejection and setbacks they faced in love.
Some other comments were enlightening as well as informative:

" ... Elise's orchestra troupe consists of her husband and brother. Justin is her husband, the orchestra's first violinist..." *Aleena Benjamin. Manipal*

"Generally, the first violinist or concertmaster is tasked to play the music (melody) while the second and others play harmony!" *Helmut Herbst, Berlin*

"...Elise was a music student of Justin. They met and married after a secret, lengthy romantic relationship, against the will of her parents!..." *anonimous*

"*Shade*!...Unfortunately, in Beethoven's case, certainly this didn't happen!" in reply from *Bergmüller, Wiesbaden, Germany*

"Hi, Gemini is the Pianist, her only brother; flutist is BalaMurali Murthy; Sojan Devan, on Mrudangham is the son of the reputed maestro Mahadevan Iyengar; the funny guy, with that *afro* haircut is Varadarajan Murthy on clarinet; Eby, the saxophonist; Sakunthala, Murthy's wife, on her Saraswathi Veena; Malini Devadas on second violin; DasMunshi Bhutt is with Tabla; Boban Thampi on trumpet, and Pius George on Trombone. Majority are alumni of the College of Music, Thiruvananthapuram" *Many Cheers!, Alumni Association, College of Music*

"... I know Elise very well, she is from Palai, my mother's hometown. Her mother Celine, was a professor of Anthropology at Loyola college, Sreekaryam" *Kamini, TVM*

"......I like Anoushka and her Sitar; both are cute & pretty beautiful!", Liam Ratcliff, Surrey

"....the band is nominated for this year's Music award", *R. Balamani, Kottayam*

"... Beethoven was born in Kőln, in my home town. I play Tabla. I Love it, Wunderbar!" *Manfred Dorfmeister, Bonn*

" Kampala, Uganda greets you, music is the universal language, second only to " bah..bah.., baby Talk"!" *Aamoo Ampala*

"...Reggae & afro Jazz my genres; this one cool too", *Kibati, Nairobi, Kenya*

" I like Bollywood dancing; they seem to change dresses frequently while dancing, haa, haa, uhm! Gareth Martins, Long Island

"..My daughter played "For Elise" for piano competition. She got the first!", *Natarajan Natar, Singapore*

"... I want to learn this and play it to my fiancé, before asking for her hand", *Rene Fonataine, Mauritius*

" *yeh, I missed it. I heard it was wonderful. I hope they will re-broadcast it"* Sasidharan Unnithan, Palakkad

...so on and so forth; the stream continues, as we speak.

Some comments were hilarious; others serious and thoughtful. Most of them were congratulatory. Besides, they talked about the concert hall itself in many words of wonder and elegance. The music sounded so beautiful that they switched on their *Dolby "surround sound'* system to enjoy it to the fullest. They tried several '*codec*'s and suggested *FLAC* gave the best listening experience. Until now, there was no surround sound audio transmission in India!

It's a matter of common knowledge that the social media often contains varied opinions & comments mixed among the facts and figures. Concealed among them, are a lot of filth, real news & "*fake news*"; gossips & rumors too. Vested interest groups and unscrupulous people, as well as foreign agents systematically hijack many good things great people have invented. They take advantage of the gullible people and vulnerable technology; divert them towards evil ends, making them impossible to use for its originally intended useful purposes. Sadly, some people systematically use the social media platforms to express veiled social, political, racial, religious spite; vilifications, innuendos, threats; and to spread hatred causing divisions among people. It is also widely used to commit crimes, frauds, scams, swindles, identity thefts, human trafficking, etc. It's a terrible shame!

The public is attracted to it for one's own social and existential recognition, and validation. Mercilessly, the engagements consume valuable time; end up dwelling on frivolous topics; hijack young and budding minds; cause serious distractions; even addiction; damages health, sight, vision; scars and subjugate fertile minds. Some users tend to cocoon themselves in their own world and see the entire world on their 11 cm dia. LCD screen which is always in their possession. Ultimately, it may simply all be just a '*fad*' or '*tulip mania*' phenomenon, marked by initial unbridled enthusiasm, followed by regression, or sudden retreat. One needs to apply critical judgment and sort out the '*goods*' and '*bads*' from what is out there on offer. One must put one's own wholesome needs, and well-being ahead of everything else!

Well, here in the hall, the opinions and instant reviews of the music helped to stoke passion to ecstatic levels among the audience. They began to demonstrate it freely, even though most of them have not heard this music piece before. Nevertheless, they seem to enjoy it tremendously, helped by the introduction of its theme at the start, as well as the numerous comments from the overseas social media sources who are most familiar with this kind of Western classical music genre. Insinuatingly, the TV director chose the shots showing coy and flirting interactions and exchanging of sly glances between Elise & Justin as the music becomes heavily expressive, and eloquent about the innermost feelings. Obviously, the *T/D* has known about the past romantic experiences of the musical couple and the sentiments contained in the music.

In the music that was played, the rhythm stayed fairly constant, indicative of the unrelenting plea for the lover's recognition and consent. In Beethoven's heart, the fire of passion has been still burning intensely, despite the lover's original *indifference* and apparent rejection. Then, there is a few '*agitative*' passages suggesting a rejection of his earnest overtures of love and the anguish caused in the aftermath.

Though the music itself was totally foreign and unheard, the sentiment of the music was quite familiar to the many in the audience.

The audience fully understood the theme and the emotional ramifications as the tune has expressed them quite vividly. Among the listeners, there is a few who suffered the same fate as Beethoven did, and for the same reason too. Many sufferings are caused by traditions, deep-rooted social prejudices, ranks & social status, customs, and practices. In the society, “*genetic matching*’ has been prevalent since the ancient times, though they have been applied and practiced by the *then known* knowledge derived from observations, statistics and deductions. Prejudicial sentiments and situations do prevail in every society and country in the world in various degrees, shapes, and forms. Pretending otherwise is feigning hypocrisy!

The design of a concert hall is based on several acoustic principles as the acoustics has a strong emotional impact and better listening experience for the listeners. The idea is to optimize the quality of listening by propagating the original sound signal in high fidelity and adequate intensity levels to the ears of every listener in the hall.

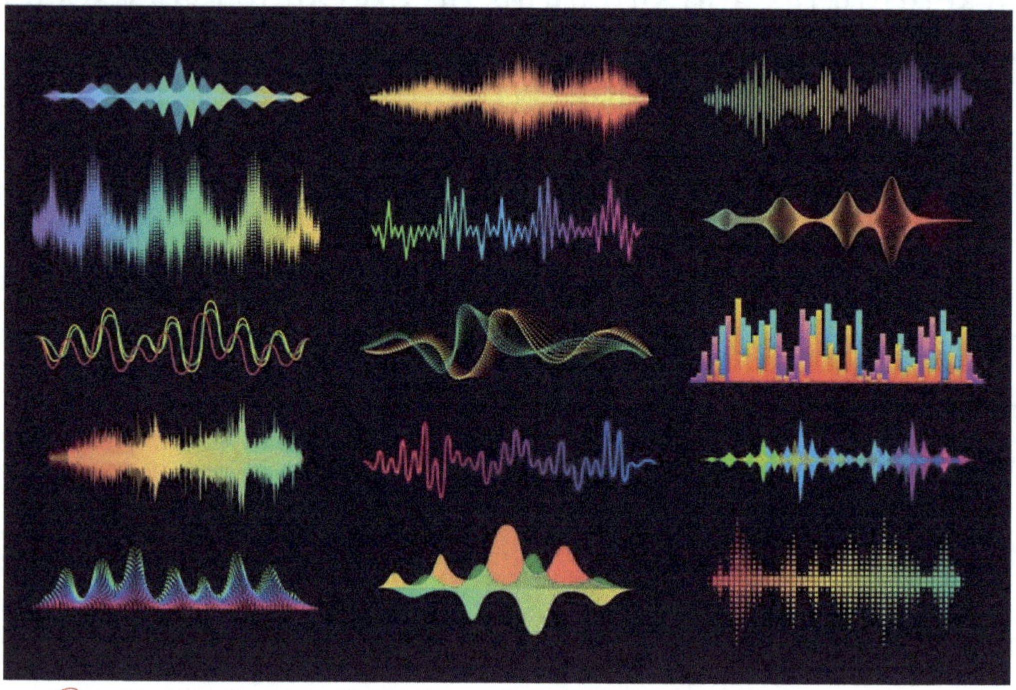

Sound waves illustrated in different forms showing properties and overtones

The audio signal itself possesses certain characteristics. It has a rather narrow wave spectrum of '*air pressure waves*' in the frequency range of 50 to 20 kHz. In spite of this perceived deficiency, it contains all the essential information needed by the senses and brains to process the auditory signal so that they can be detected as a '*hearing sensation*' which we call '*sound*'.

By this time, we learned that the sound signal is a wave of vibrations /disturbances in the medium unlike *electromagnetic* radiations. The vibrations travel through a medium of matter containing microparticles and by the phenomenon of compression &

expansion transferring energy from one to the adjacent particles in the 3D space. It must be noted that the greatest audio sensitivity and hence the pleasure of music is felt in the stingily narrow range of 80 Hz, to 2.4 kHz, and the "*sweet spot*" (a term borrowed from ball & bat /racket games) is just within 220 to 1.4 kHz. The sound waves propagate as wavefronts and travels through a medium around us at an average sonic speed of about 330 m/sec. In a closed space such as a concert hall, the waves are subjected to various modifications, reverberations, attenuations, distortions (together they may be termed '*impairments*'), and even sometimes amplification due to the fact that they may bounce around multiple times and pass through resonant objects. The sound may suffer absorption, reflections by materials with different natural properties. The waves may be reflected, deviated from its original paths, split, separated and therefore arrive at a certain point at slightly different times. The waves may '*excite*' molecules of other materials as it impinges on them, causing it to resonate in its own intrinsic oscillations, adding noise, latency (time interval), and distortions. The resulting effects are several folds: such as echo, reverberation, garbled sound, noise, and attenuation of the signal. Some effects may be still pleasant while others are too annoying and affecting clarity. The medium of propagation has various properties too, such as temperature, humidity, air pressure, absorption characteristics. To reproduce the sound in its original form and integrity as it arrives at the listener's ear is therefore an important consideration, despite being a daunting task. These factors bring about the significance of acoustic design in concert halls and theaters.

The *audiophiles* (people seeking and enjoying the best quality audio) and audio enthusiasts consider these matters when buying and setting up Hi-Fi sound systems for home, living room, hall, meeting, conference rooms, etc. Hi-Fi, basically, refers to the high quality audio reproduction over the full range of the audio frequency spectrum true to the original (studio recorded sound under controlled acoustic conditions) signal. They have very high technical specifications of different audio parameters, such as bandwidth, intelligibility, low noise, distortion, etc. Besides, individual parameters can be controlled and set to personal taste or respective to room acoustics. Hi-Fi is mostly associated with purely audio transmission, or audio-only sources, such as tapes, CD, FM radio transmissions, audio streaming, etc. The audio controls for a PA system (Public Address) is minimal, a volume control, possibly separate bass & treble knobs because it is propagated to *uncontained* open space where the elements and conditions cannot be controlled.

In order to achieve Hi-Fi & Dolby quality, a large amount of data bits need to be processed in real time in great speed and within stringent error limits. When every constituent factor is addressed properly, it demands an even larger bandwidth of frequencies to transmit the whole amount of information. A clever and practical solution is to encode the signal by digital processing. Most everything and every bit stream such as L & R signals of a stereo, 7.1 (the '7' denotes the number of speakers while the '1' the number of subwoofers) channels of surround sound are processed, equalized, amplified, and encoded. A number of mathematical operations are applied in the process, such as '*Laplace transform*' used in analyzing the dynamic control systems. Laplace transform changes one signal into another according to some fixed set of rules or formulas. The repeated information (redundant) streams are filtered out. Only a representative *replica* signal is packaged, reconstructed, and reinserted at the receiving end. Audio '*codec*'s that encodes & decodes using various schemes are used to achieve this task. Most widely used formats are software apps such as *mp3, wav, aac, vorbis, & flac, etc*. There are hardware based codecs too! Interoperability, meaning its ability to automatically detect, encode, and decode different formats, application situations with as many brands of equipment are critical considerations in choosing audio & video codecs.

Thus, a stereo has a discrete left and a separate right channel. Then, there is 7.1 surround system that carries 7 discrete channels, mainly intended for '*direction location*' namely, Front right & Left; Center; Surround right & Surround Left; Surround Rear right & Surround Rear Left; plus a Low Frequency Enhancement (LFE, widely known as **Sub woofer**, producing extreme bass range sound). Each channel is delivered to the respective Loudspeaker only. Mp3 codec supports 5.1 with one LFE channel while *AAC* (Advanced Audio Coding) supports 16. Vorbis can support 255 discreet channels including mono, stereo, quadraphonic, polyphonic, 5.1, 7.1, 9.1 Dolby surround sound, etc. These additional capacity is very useful to transmit multiple languages on their own specific channels such as in International conferences /sports commentaries on satellite channels, as well as descriptive audio for "*visually impaired*' participants. The format has become the *de facto* choice for Blu-ray disc and HDTV (Hi-Definition TV).

To begin with, the Art Center uses AAC format helping to carry instrument and voice specific channels in the auditorium. Another term popularly used in this context is "*ambisonics*" meaning '*a form of surround-sound system that attempts to reproduce the directional properties of the sound adding extra dimensions of realism and depth of sound*'. Undoubtedly, it enhances the immersive experience of the listeners & spectators by providing the ability to locate and identify the sound sources. One could sense the aeroplane flying overhead as the sound comes from an elevated location; hear which way the footsteps are approaching or leaving; the train '*swishing*' past an onlooker, and so on.

All audio streams can be detected by its digital '*ID flag*' code helping to properly decode information automagically needing no operator intervention whatsoever. In the digital world, form & formats change frequently. When one or more of the related things change, the other devices in its chain require further revisions and enhancements. Most are accomplished by software updates, or generally called '*plug-ins*'. If that's not adequate, new hardware will be needed, and probably with new generation of input output ports; terminals and cables. Analog to digital transformation is one of them. They may sometimes make old technology (legacy technology) and devices inefficient, inoperable lacking features, *intelligence*, quality, or compatibility; thus, simply making them totally obsolete and merely junk.

Here, in the concert hall, when a music, or other performances are staged, they are accompanied /rendered /packaged with visual image, video, animation, text, and data information. It becomes necessary to accommodate additional foreground, background, and respective soundtracks distinctive to each of the performance. In such cases, "Hi-Def" (high definition) technology is employed to match with the 'HD' (Hi-Def) Video. This is a hi-quality *digital audio* made to complement the HD video materials. Since there are many algorithms and standards, the selected standard at the concert hall must be seamlessly compatible to all the digital transmission and recording standards commonly used. Thankfully, the digital broadcast standard adequately address this requirement. Most of the streaming networks, social media channels, as well as conferencing (video calls, Tele-conferencing, tele-medicine, remote hybrid work) providers ensure such device compatibility. If not, their products and services will not find a marketplace.

Furthermore, there are many sound effects & expressions created and produced by the artists, performers, and instruments, or even externally which need to be fully integrated. In drama, opera, dance, movies, and other performances, they include so many other sounds from the nature, animals, machines, surroundings and so on, besides background music scores. The challenge in a good concert hall is to carry any sound in the *human audio range* faithfully and seamlessly throughout the facility. In music, the usual sound effects comprise ornamentations, articulations, vibrato, modulation, resonance, attack, decay, accent, timbre, and various playing styles specific to instruments and musicians. Some environmental factors can affect the audibility and quality of these effects; some effects are extremely delicate, feeble, transient, and therefore more sensitive than others. No matter what, irrespective of the different characteristics and qualities of the sounds, obviously, they can occur only in the human hearing range. That's where all the focus and energy must be expended!

This new and acoustically designed concert hall has a large dome. Besides being architectural and artistic in nature, it is designed to produce the '*cathedral effect,*' a unique, natural reverberation, and *sustained* sound effect. The dome sits on a colonnade of 64 equally spaced columns. The columns are smooth and cylindrical in shape to help deflect and distribute sound signals smoothly and equally in all directions. The dome is *centered over the stage*, but from the level ground, appears to be sitting in the middle of the marvelous edifice. There is an array of loudspeaker system with an integrated *'AI'* (Artificial Intelligence) module & *machine learning*, which analyze the sound spectrum in real time and beam the sound to the spherical surface of the dome which is equipped with flexible, serrated acoustic baffles that disperse the sound beamed towards the audience. *The sound fills the hall*! It comes to the ears of the listeners *'measuredly'* (in small measurement or increment) milliseconds delayed, giving another dimension in 3D (full sphere surround sound) hearing. A church *pipe organ* is the perfect instrument to bring about the particular sound sensation to the fullest effect. This is because of the fact, that the pipes are arranged in *rank*s with common *timbre and volume*; *stop*s of many ranks of pipes of differing *timbre, pitch, and volume*, or simply many sounds of *unison* and *polyphony* tones. In this hall, there is no organ pipes installed. The effect can only be artificially produced and used only experimentally. It's reserved for special occasion until its usability and proper applications are fully vetted and realized.

Again, embedded on the walls are thin flat panel loudspeakers, devoid of magnets or large diaphragm. The entire panel vibrates like the piano soundboard to create sound waves. Each *'pixel'* in the panel vibrates at a resonant audio frequency; an array of such individual pixels amplifies and reproduces sound in hi-fidelity, and controlled intensity. They are the *'state-of-the art'* sound reproduction components covered and decorated with famous paintings of Mughal, Madhubani, Raja Ravi Varma, Abanindranath Tagore, and from famous ancient Kerala arts. In another section of the wall there are hi-resolution, panoramic photographic images of Kalaripayatt (Kerala martial arts) & "*Kadhakali*" (a uniquely Keralite 'Pantomime', a play, dance, and theatrical performance characterized by wordless storytelling, using only bodily movements, gestures, and facial expressions) images. They may look like wall paper, but really are thinly imprinted technology capable of displaying the objects moving in *virtual 3D*. The panels propagate direct sound, over a fair distance as a beam of sound waves, without reflection or degradation of volume level, avoiding feedback, reaching the ears of the listeners in direct straight-line path.

Another distinct feature of the performance hall is the HD (High Definition) Dolby surround sound system. It delivers sound in 3D dimensional or spherical envelope- from every side, behind, above and even below. It creates a natural, multidimensional, immersive (wholly / deeply engaging effect) sonic environment, based on the science of the natural human hearing and sensation. In other words, it supports the human ability to precisely locate the origin, direction, and strength of sound. Furthermore, it enhances the overall experience of hearing all components, harmonics, and delicate transitions, providing also a real sense of the individual instrument and voices which otherwise fail to register in listeners' minds. The listener can distinctly discern the source and location of the incoming sound. Normally, the piano is located on the left, violin nearby to the right, drums from the back, woodwind in the middle, the *consorts* (ensemble of the same musical instrument family) of string families in the front row; the blaring brass instruments further back, but in front of the loud drums and the percussions (drums, kettle drums(timpani),cymbals).

A symphonic Orchestra generally, consists of piano; families/groups of Violin, Cello, Wind, Woodwind, Brass, Percussions; and sometimes a harp, as well as a choir. Though some are not commonly used in Indian music, accommodations are made to use almost any instruments when foreign troupes visit for guest or cultural performances, and for international festivals.

In Indian music, the rhythm ensemble of Tabla, drums, and other percussions group play squatting down flat on the surface; so also traditional Karnatic violinists. To be visible, and for the musical reasons, they have to sit at the front; tabla onto the far left and violin in a front-central locations. The woodwinds of Nadaswaram group also play in the same manner. Its sound is loud and shrill; therefore, moved to a hind row on an elevated platform.

Life, dance, music, and divinity exist together. The musical instruments accompany and enhance them all: They are complementary and mutually stimulating!

- A word on String instruments

Every instrument has its own tonal quality, voice range, harmonics, and other subtle, but *listener-discernible* characteristics.

The acoustic string instruments are delicate, and the best ones are made by special craftsmen, '*luthiers*'. The violin is the most prominent and an integral part of the philharmonic orchestra. It's the largest instrument group consisting about the half of all instruments. The overwhelming number of classical compositions are written for violin. The story goes that Mozart was rebuked by his father and music teacher Leopold that he is interested in composing only violin music. Apparently, Mozart took it quite personally, may be even disparagingly. Within the next fortnight, he composed 4 piano sonatas and handed over to his father. They had been some of the most acclaimed works ever since, among his prolific collections!

The other categories of instruments accompany its lead and embellish the harmony, and fill the overall sound spectrum. Thus, violins earn the most prolific of all the strings in any symphony orchestra. It is a string instrument with E, A, D, G, string pattern; has a frequency range of 190 Hz - 2.4 kHz, dominantly finding its presence in the '*sublime sweet spot*' of human listening sensation, perhaps explaining human affinity to its sound. Besides, it plays notes with excellent resonance, *vibrato, tremolo, sustain* effects in agile, versatile, responsive ways, and in sweet *unison* (identity of pitch). Its sound blends, accommodates and resonates well with all other orchestral instruments. The quality of the sound is determined by the resonant qualities of the selected close-grained wood material, aging of the violin, the quality of the strings, and the friction produced between the bow hair and strings that could be controlled by using *rosin*. Violins when bowed produce *fundamental* and '*overtones*'. As the vibration energy lingers on the strings a longer period of time, they produce sustained pitch.

A symphonic orchestra conductor is required to know how to play the violin and piano, the two most important pieces of instruments in the modern symphony. There are a few conductors of great orchestras who lack such specific experience. One must assume that they have distinguished sound sensing ability and special talents to overcome the apparent deficiency.

The *viola* is slightly larger than violin and like violin is a 4 stringed instruments, only that it has a string pattern of C, G, D, A, reaching down to the C note, a 5th note lower than the violin. It has an approximate frequency range of 130 Hz - 1.8 kHz. As it lies in the range of clarinet and bassoon, it forms a special tonal connection and association with them.

The *cello*, like viola has the same string pattern, but pitched one full octave lower with heavier strings. It is much larger, consequently held slanting on to the body, supported on the floor by a *Spike* located at its base. Essentially, it is played in similar bowing and plucking actions. The qualification the "*dulcet tones of the cello*" best describes the sweetness of its sound. It sounds serious, sombre, and aloof. The lowest note is pitched at *C2*.

The *double-bass*, or simply *bass* is the lowest sounding of all string instruments; plays in the extreme low range of the sound spectrum. It has a frequency range of about 45–390 Hz, remarkably of course, almost a cool 4 octave range! It forms the basis of rhythm and rhythmic harmonics and provides the distinct bass rhythm for the orchestra. It is almost twice the size of a cello which is also the second-largest string instrument, deriving its qualification for its name '*double bass*'. It has similar stringing pattern as a violin, E, A , D , G , but 2 octaves down in the very lowest part of human hearing range; obviously, have heavier and longer strings. It can be played by a bow (arco) or by plucking (pizzicato) as the jazz musicians prefer to do. Sometimes it sounds like the lion growling; the playing movements of the fingers as a restive animal like the monkeys jumping up and down a pole. Please bear in mind that music is not a circus which involves physical movements with agility and finesse appealing mainly to visual senses. Music is an exercise of sound agility, ranges, and finesse appealing to the auditory senses, sometimes inducing movements of body and limbs; also increasing blood flow to the brain.

- ***A word on woodwind instruments***

The woodwinds provide pleasantly contrasting timbres. The *flute* (Murali), the best known member of the woodwind family is a popular instrument of every Indian and western orchestral music. Flutes made out of animal bones were the oldest instruments played. The shepherds are pictured with playing it, possibly because its soothing tone relates well to loneliness, nostalgia, and the elements of the nature. The sheep likes it too; helping them chew the grass well and grow faster!

The flute is used in solo, group or orchestral settings. It has an impressive Frequency Range 261 Hz - 2.4 kHz. They can be made at suitable lengths to produce the required range. It is the highest pitched wood instrument, but lower than the Piccolo of the woodwind section and blends smoothly with other orchestral sounds. It is made from many materials- wood, metal, and bamboo. The wooden flutes are very popular with Indian folks musicians. The Far Eastern folks prefer the bamboo made. Naturally, the timbre of sound varies with the material it's made of as the particles vibrate differently. The flutes are the easiest instrument to craft.

In India, Tibet, and perhaps in many Buddhist's '*Dharmasalas*' around East Asia, *conch shells* or the '*Sankhs*' are used even today in chants and devotional music. These ancient and natural instruments of more than 15,000 years old produce deep, penetrating, and plaintive sound. Perhaps, the specie of sea creatures from which the shells were taken from, had become extinct some time ago! Basically, the *sankh* is a '*cavity resonator*' and will have a resonant frequency purely dependent on its size and shape. The marine *Foghorn* (a mechanical instrument sounded at intervals to serve as a warning to vessels in fog) may be the only other similar sounding instrument.

The *clarinet* has a single reed, is nearly cylindrical, closed at one end and open at the other. Applying '*Acoustic Physics*', the so-shaped instruments can only produce *odd* harmonics with 3, 5, 7, times the initial *fundamental frequency*.

The *oboe* has double reeds; one reed similar to Clarinet; the other tuned to the next higher octave. As to be expected, it makes a mellow sound with the addition of the higher pitch. In all other respects and harmonics, it is similar to the clarinet though distinguished by a different timber.

Piccolo, is a half-sized flute and plays an octave higher. At the highest note, it sounds shrill to many. The fingering is identical to that of the flute. Flute players often alternate with piccolo when playing the highest passages. It has a Frequency Range of 600 Hz - 4.2 kHz. The highest note the last key on the 88-key grand piano, $C^{8.}$ also the same.

The *Bassoon* is the lowest sounding woodwind instrument; has *double* reeds. Its sounds identical to that of a *baritone* singer with a quivering resonance, a typical characteristic of two sound producing sources working in tandem. It is a long narrow tubular instrument and has a Frequency Range of 60 – 590 Hz. The fingering is very similar in playing most of this category of instruments; consequently all of them can be mastered by the same instrumentalist!

- A word on Brass instruments

The brass provides increased vitality, vigor, and power. Being made of metal, they produce sonorous sound. This category includes the Horns, Trumpets, Trombones, Tubas, and variations of them. All

among this class of instruments are popular and prominent in Jazz, Dixie, Parades, as well as Alpine music.
The Horns have a weaker sound than the rest of the brass instruments; it resonates well with the woodwind instruments.

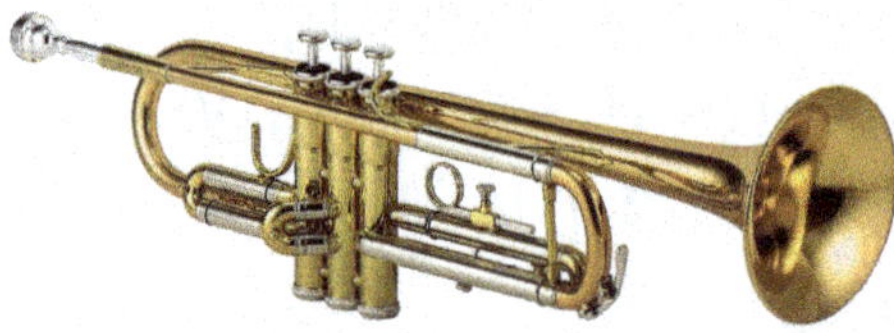

The *Trumpet* looks like twisted, bend, coiled pieces of 4 cylindrical tubes; 3 in the linear direction and a tiny piece in the transverse. Apart from the main long tube, there are two other tubes of varying lengths which alters the notes by certain degree of steps. These tubes are switched on by valves operated by the three fingers to select any additional tube. The mouthpiece can be pulled out or inserted. It is the actual sound generator! The sound goes in to the tube and comes out of the *Bell* at the other end, powerfully ejecting the sound outwards. The long tube enhances and amplifies the sound and gives its tonal quality with its character of '*internal reflections*'. The tension caused by the lip movements by '*buzzing*' into it produces changing vibrations, generating notes of the pitch, and other effects. The Jazz trumpeters are often seen with bulging cheeks, while blowing the trumpet much like that of squirrels, when they gather nuts and food items in the mouth to take them away to hide for the future use. There is always something for us to learn from the simple creatures ubiquitously seen around. Besides, the squirrels quietly teach the value of "*prudence & providence*"!

On the trumpet, the 2nd Valve opens an entrance port to the tiny piece of tube to alter the note lower by a '*Semitone*' (1/2 step); the 1st valve opens the port of a tube to lower the tone by a '*Full tone*' (full step); the 3rd valve opens the port for the longer tube which lowers the pitch by a '*Minor third*'. The three valves with its different combinations 1, 2, 3;+(1&2); (2&3); (1&3); (1&2&3) (as shown by the 'combinations') can produce the 7 chromatic notes. By modifying the way the valves are pressed, a trumpet player can add '*color*' to the notes. They have a Frequency Range of 160 Hz - 1.0 kHz.
The *Cornet* is its close cousin.

Trombones, looks like an extra-long trumpet; is the longest in the brass family. It's an indispensable instrument in Dixie, Jazz, swing music. Its '**stop & go**' style is unique! They are made in alto, tenor, and bass ranges; has Frequency Range 80 – 520 Hz. It is played by moving a slide. It can slide through notes smoothly. So playing a '*glissando*' ornamentation comes to it rather naturally. Watch out! Keep a distance from it; when extended, it may hit you, and the music get distorted. Everyone will notice it; you get laughed at!

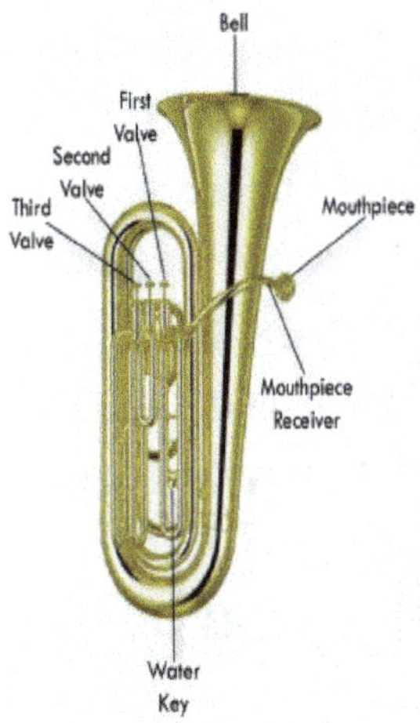

The *Tuba* is the largest of the brass family of musical instruments; applies the same basic principle and played much like a trumpet; It has a 4th valve to enable to play very low sound; has a rich bass sound, therefore, has the longest resonant air column. The instrument is kept compact by making the long tube in the shape of several loops terminating in to a progressively bulging Bell, like a reticulated (marked with lines that resemble a network) python snake, wound around a prey constricting it. The frequency Range of tubas is 40 – 260 Hz, and plays the lowest passage of the Bass clef. At the lower range, it sounds like a grump of animal snorting "**Phoom...Phoom**..."It can create '*swells*', '*punctuations*,' etc. to accentuate stress or emphasis. It is also used for playing '*Baritone*' melodies. It is played by holding the instrument close to the body and held at an oblique but comfortable angle; the bell, thankfully ejecting sound upwards, sparing jolting anyone off in its direct path. (The sound emerging out of it, has high energy, and may be damaging to the eardrum!)

- A word on the percussions

This category includes all kind of drums, struck by hand, palm and fingers, and / or drumsticks. The tablas, Chenda, Mrudangam, and Thavil, 'Dhol, Dholak, Dimdi, Ghatam'; Congas, Bongos, tubular bells, Xylophone; membranophone, vibraphone, idiophone; anything that produces short periods of sounds by striking, plucking, or any '*impulse*' (Phy: force acting for a short period of time) forces belong to this class. They form the rhythm making group, include both tuned or non-tuned pieces. A music without rhythm looses and lacks more than half of the music. I feel the frustration when I try playing Indian film songs without the tabla accompaniment. The accordion bass sounds feel so inadequate!

- A word on the Harp

The *harp* has 47 strings, 7 pedals, 3 on the left for D,C,B; and 4 on the right for E,F,G,A. Its string pattern has an identical shape to those of the piano. It's played holding vertically. It has wide and ancient roots; seen in many cultures and geographical regions, albeit in different shape, modes, and styles. The Harp uses colored strings of nylon, gut, synthetic gut (as similar in Tennis rackets), or wire to identify notes; Red for C; Black as F; White as E, G, A & B. The Harp is played by a total of 8 fingers 4 from each hand, except the tiny fingers. The 7 pedals, corresponding to the 7 keynotes, played by both feet are used to play the Flats (b) or the sharps (#) of the keynotes.

It has a legendary significance in Jewish beliefs. King Saul, had been constantly afflicted with nightmares, anger, fits, and general malaise. That affected his mind and body in debilitating ways. David, from the House of David, who also later became King, was brought in to play the harp and sooth and mellow the emotions of the king. As he was playing, King Saul began to feel progressively better; the evil spirit departed from Saul. It was the earliest musical '*exorcism*" of sorts. The king was cured by the music! There were no tablets or IV solution in those days! (soothsayers who deal in the interpretations of dreams were the only other option).

The therapeutic, soothing and relaxing qualities of Harp music is practiced in the religion since that time. In general, music is believed to have several beneficial effects on relaxation, sleep improvement, mitigating sensations of pain, anxiety, and restlessness; stabilizes vital signs; improves the mood of the listener, as well as the blood flow to the brains. So much so, harp music is played in the '*end-of-life vigil*' to help the souls in their final moments for the departure to the blessed land of eternal '*afterlife*' in an as fittingly tranquil, and peaceful manner as possible. The faithfuls believe that the music helps the serene separation of the soul from the body; escorts it right up to the '*gates of heaven*'. As the spirit reaches the destination in good stead and spirit, the entry to the '*heavenly kingdom*' is best and rest assured!

Ancient civilizations in South Asia had also once used instruments of the Harp family. The equivalent of harp in Arabic music is '*Qanum*', trapezoid-shaped flat board over which 81 tuned strings are tightly strung. In that sense, it is more similar to the piano. The gypsies also use similar instruments played by a '*striker*' in each hand.

A modern concert hall must accommodate and faithfully reproduce the entire *gamut* (complete range; the entire series of recognized notes) of individual sound bytes including all the audible overtones.

When we get in to the bathroom for a bath, we tend to sing and quickly discover that we have some hidden singing talents as shower falls on the body by force of gravity. It gently drains down from head to toe, hugging the contour of the shapely body; releases its potential energy to every pore it touches; open them up; increases the oxygen intake. The falling water gives a soft massage and soothes the entire aching and tired body. Now, the whole body, significantly, the throat / larynx, (an organ of the neck involved in ***breath control***, protection of the *trachea* and ***sound production***, housing the '***vocal cords***'); *torso* (The main frame of the human body from the neck to groin, protecting vital internal organs, heart, lungs, lever, kidneys, stomach, intestines, etc) become damp; soft and tender. They are energized, agile, excited, lubricated, and primed; fit for effortless motility! The chest is relaxed; they expand and contracts normally without feeling heavy; hearts pump without undue labor; the diaphragm (a dome shaped muscle used mainly in respiration and other related functions) relaxes. We feel relaxed, energetic, and lively. We're in a little dreamy world of our own. Inhibitions wane. No one is around. There's no awareness of stage fright. When we murmur, speak, or sing the sound emanates as vibrant, smooth, and sweet. The sound reverberates and comes back to us with amplified energy. We feel a pleasant sensation and a deeper perception of our own voice that comes straight back to our ears and stimulate minds. Further, the surrounding air becomes moist and heavier. The moisture diffuses in the space within the narrow walls. The ambience, situations, and conditions are ideal now and responsive for optimum voice production and hearing. Spontaneously, we sing; we like what we're hearing; continue the same and become very pleased with it. We become confident; suddenly, become singers! It's so wonderful!

The sound feels sweet, vibrant, and enchanting because of the excellent *acoustics* of the environment and the space within which the voice spreads. Within this enclosure, the sound reflects off the closely surrounding walls. It gathers reverberations and travels faster through the moist air reaching our ears almost instantly. The hearing and sound perception elevate themselves to higher levels. The humidity is continuously maintained by the falling water. Within the bathroom space, there is no major objects to absorb or deflect the sound energy. The glazed walls only reinforce the sound waves without distortion. For all intend and purposes, it's an ideal, acoustic, Hi-Fi environment! If we speak /sing in immediate proximity to the corners of the 2 walls, we can feel the sound responds with a deeper '*bass-range*' quality. We can vary the distance and feel the pronounced changes in the sound we hear. Further, the sound feels louder for two main reasons. There is little attenuation of the sound signal and the volume of the space is rather small; the sound fills the space. Besides, the sound fades only gradually. Technically, everything helps to maintain a very high '*Signal to Noise ratio*' without annoying echo, noise, or distortion.

But if one performs the same way in a thatched bathroom enclosure, the results are not that flattering at all. The dry leaves simply absorb almost all the sound energy leaving nothing to reflect; the singer gets no feedback. The sound has no resonance, reduced sound levels, and feels dull with no *musicality,* just as *thud*. We immediately realize, it's just a covered place for a quick, quite bath, and not meant for singing. We are beginning to recognize the importance of acoustics in an environment used for communication, and that each of such places has its own optimum set of requirements for the best audibility.

Theoretically, we could build any such acoustic enclosure using mostly reflective materials and modify the character of sound within that environment in artificial ways. There are also many electronic methods the recording artists use in '*live*' as well as post-production situations. The foregone simple analogy is to emphasize the vital role of *acoustics* and what should be desired in auditorium, concert hall, theater, and the like.

I still entertain certain faint childhood recollections of my multi-talented uncle routinely waking up an hour or so before the sunrise and going for a bath in a remote canal (those days, there were no 'attached' bathrooms nor toilet amenities inside houses. It was not considered hygienic nor desirably private!). He used to sing loud to the top of his voice; perform various voice exercises, besides practicing swimming. At the time of early morning, water is pleasantly lukewarm (at about 2 m below ground level, the temperature is almost constant at about 22°C, the principle applied in geothermal cooling systems), the ambience is quite; the thrust of water pressure provides a counterbalance to the soft movements of the lungs, increasing its strength, volume, and control; the diaphragm, a primary organ in sound production (besides serving the main functions in respiration, coughing, sneezing, etc.) becomes relaxed; the throat feels cool, (whereas, friction retards motion; produces heat, adversely affecting efficiency; a reason why we use lubricating oil in engines!) loosening up throat muscles which control vocal cords. He discovered a truly natural setting where music can be practiced and enjoyed better. His concluding song was always a devotional '*Heliolatry*'. Some people far across the banks said that always did they wake up to his classical '*serenade*' in great cheer and devotion to '*sun-god*'. A few of them regularly joined him in their morning sun worship rituals! His rather short life deprived us of learning from his great insights, deep wisdom, and discipline. In his honor, memory, and as a tribute, I wrote a heliolatry music titled "*SooryaNamaskar*" and posted on my website and YouTube: SooryaNamaskar (Heliolatry-sun worship) .

The space and the medium through which the sound waves travel is of critical importance. It is the task of architectural engineers to provide a suitable design, meeting all the acoustical requirements. The structural engineers shall build it accordingly. A variety of specialty soundproofing and noise control products and materials are used to achieve the desired acoustic quality.

Soundproofing is done to prevent exterior noise entering the hall that includes the backstage noise as well. Usually quilted curtains with sound absorbing quality is used at this location, still allowing frequent traffic of actors, performers, instruments, stage crew, and stage scene sets. Another item is '*echo eliminator*', which is a high performance noise control cotton product. Then, there are vinyl acoustic wall panels with frequency specific '*sound absorption coefficients*' that can be conveniently hung on the wall. Certain audio artifacts could be '*equalized*' by electronic means and audio system components. But over-reliance on them must be avoided. All that can be done should be done at the design & construction stage. In a large open space, the signal suffer many kinds of impairments: noise, *attenuation* (gradual diminishing of signal strength) multi-path reflections, absorption, echo, and so on. In concert halls, the signal has to be amplified in its true form and range to sufficient power levels so that it can be naturally and comfortably listened to. It's mainly a function of the electronic sound amplifiers, and associated equipment. Equally important components of a sound system are the Microphones, because of the fact it is the first in the chain of sound system that provides input signal. Loudspeakers are the final delivery devise and must conform to the applied specifications, and properties. The location and installations of loudspeakers are also so very important, including positioning, cabling, etc. The sound has to be distributed evenly without any '*dead spots*' or '*hot spots*'. The sound characteristics or the acoustics of the halls alter widely with the presence of audience. The human body absorbs sound; some body parts and limbs deflect; most do both. The ambient noise increases. The breathing, talking, movements, and rustles add more disturbances; the ambient temperature also increases which has a significant effect on sound wave propagation. Many of them cannot be compensated, though can be mitigated. The HVAC (Heating Ventilation & A/C) environmental conditioning system must be designed to run as quite as possible; gently, smoothly without '*surges*', '*streaks*' or leaks. The sound waves travel in a medium in alternate compression and expansion cycles from the source of sound. The medium contains particles that transfer energy to the next one, thus propagating in the three-dimensional space spreading energy. The sound travels fastest in solids than in liquids; faster in liquids than in gases. Not surprisingly, vacuum cannot conduct sound waves at all, as it is devoid of matter! The speed of sound is affected by *temperature* (as it affects the '*elastic quality*' and the energy of the medium), humidity (speed of sound increases with humidity, as the humidity makes the air denser), density (higher the density faster the sound, as the molecules are densely packed and transfer energy faster). In auditoriums air is the medium. Therefore, it's a very important constituent to be treated with certain special considerations with respect to the factors discussed above. Humidity control that should always remain subservient to health matters, figure very prominently in this context.

In typical concert hall stages, many wooden panels can be noticed. The piano side wall is lined with individual long, comparatively thicker, while the Violin side with thinner, shorter wooden panels with solid or veneered finishes; linear, or in grid arrangements, and with different perforation options for the desired acoustic performance. The wood selections offer natural looks, and ample flexibility. They are so selected to satisfy the proportional relationships and integrity of '*wavelengths*' (the length of a wave has an inverse relationship to frequency; the **longer** the wavelength, the **lower** the frequency, or cycles/sec. The note ***A1*** has a frequency of 55 Hz or cycles/sec, and wavelength of 627.27 cm!) of the frequencies the instruments naturally produce. They must resonate and reproduce the whole sound in real time and in true *fidelity*! This shows how critical is the consideration for each fitting, object, component, and accessory in a concert hall.

Let's now return to the song: '*Für Elise*'!

This *is* the first time ever the audience were able to experience the enriched and enhanced sound, in *hi -fi* (High Fidelity) & HD (High-Def) and surround sound that is enjoyable from any seat or location. *They are available on separate channels and is 'user-selectable'* on the headphone provided at every seat. All listeners can certainly hear the sound via the theater sound system. They have already sensed and felt the distinct difference and immensely enjoyed the *discernment* in taste and qualities of the sound.

Under the wonderful settings of the new acoustical and technical features, the song became uniquely vibrant and melodious. The stimulating and embellished feedback reach the stage as well. The performers instantly receive the cue (a signal, used to prompt another event in a performance) and the spontaneous reactions from the audience. They respond accordingly; play even better than they could ever have done before.

The eurhythmic (characterized by a pleasing rhythm; harmoniously ordered & proportioned) music '*Für Elise*' ended in suspenseful and emotionally captivating concluding passage which gradually faded in to a lulling triple pianissimo (ppp- played as

softly as possible ending in a suspenseful, prolonged silence) indicating heart-wrenching sadness and painful *heartache* with long lasting sorrow.
Poor Ludwig Van Beethoven!
My heartfelt sympathies to you, *Herr maestro!* (dear, great maestro!)

Continuous applause followed, sympathizing with the sad sentiment of the tune and the culminating emotional ending. Many think that it is the most captivating and powerful *love song* they have ever heard. There are many among them with broken hearts and unfulfilled love. They re-discover a new meaning and expression of their own feelings. Many are still grieving; may someday be able to recite it to the aggrieved person, given a chance, and occasion. Others may seal them secretly, tight in their hearts for the rest of their lives. At the conclusion of the song, the exact melancholic sentiment prevails in the hall.

Yet, encore applauses break out sporadically; starting a new round soon after the former fades in silence. The listeners are now emotionally involved to the core with the score they just heard. It touched everyone's tender-most emotions of love and despair.
All in all, it's an overwhelming scene; a fitting reaction to a rather common sentiment in the society!

The audience could not possibly go away without hearing more of the same, or similar kind of music. They are emotionally entranced; brought to a '*point of no return*'!
Indeed, they want more! Many requests follow by the soundings and pleas for yet another song.

They shout "*please let's have one more!... just one more*!",
like kids pleading mothers for cookies in a candy store.

Nobody could possibly ignore the fervor and the sincerity of the audience.
The musicians are in a dilemma. They can't simply walk away. They don't dare to disappoint the listeners. The '*cameo'* musicians are also

experiencing the excitement of their own performance. In this sonorous and magnificent hall, the music sounded extraordinarily bright and brilliant with delectable embellishments and refined replication.

Everything may pass! But, in the hall, at this moment, nothing else matters! Everyone is, literally, at the climatic moment of enjoyment. They have never so genuinely and thoroughly savored a concert like this ever before. Their pleas are impassioned and devout!

During this period, Mrs. is seen coming from behind her husband and speaking into his ears. He checks his watch and nods in tacit consent. The '*trio*' musicians are also drawn into the prevailing frenzy. They are completely enthralled and absorbed in the spirit of the moment, forgetting other extremely important things they are supposed to be doing in the very next hour. They have never played in such an acoustically designed, detailed, appointed, and responsive concert hall. They hear their own music in splendid tone and timbre; resonance, and reverberation. It has been a pleasantly unique experience, that may almost surely never come by again. They have their studies, careers, and occupation. Music is for them, just an avid hobby and passion; a family bonding exercise to keep them together, happy, and well. It was never intended for becoming stage performers in any stretch of the imagination.
For them, this is a wonderful occasion to cherish and savor for ever! They are very much inclined to relish yet another *go* at it!

When the applause subsided, the accordionist announces his family's last song.

"Well, ladies and gentlemen! For our very last song, we choose a magnificent Wedding waltz. It is universally known as The '*Second Waltz*', by a Russian composer,... Shostakovich. The popularity of this song soared after it was played for a ballroom dance scene in the grandiose settings in the classic movie "*Anna Karenina*" based on a novel by Leo Tolstoy.

The drama, *Anna Karenina* narrates a sensationally romantic tragedy in the rich world of the opulent Czarist times; Anna's extramarital adventures & escapades; ensuing scandalous drama; reckless and lavish lifestyles of the elite in stark contrast to the deprivations and wretchedness endured by the Russian farming villagers; an earnest campaign by a lone idealistic landowner; the widely divergent attitudes of many leading characters; capricious passions and trivial pursuits of the contrasting social classes in many moving details and in huge volume of folios. These themes and accounts as well as those in "*War and peace*" were at that time the most voluminous of all novels!

The Nobel Committee for Literature at the Swedish Academy received intense criticism for **not** awarding the inaugural, *Nobel Prize for Literatur*e to Tolstoy in 1901. He was thought to be the '*most revered patriarch of literature*' of the time, besides being the '*greatest and most profound poet*. His novel, War and Peace had garnered worldwide acclaim and elevated him as the obvious choice. The decision was sabotaged when the committee yielded to the incessant public relations campaign, pressure, and endorsement by the French Academy for Sully Prudhomme who was berated and ridiculed as '*Tolstoy's inferior*' and was, as soon forgotten. As a consummate socialist and pacifist, Tolstoy quipped: "It spared me of a huge problem of how to use this money; that much money can only bring evil!", a philosophy diametrically opposite to that of the billionaires.

Thereafter, Leo Tolstoy never coveted the prestigious award; took himself off from any further reckoning and consideration for it in the subsequent years of his life.

Besides being all the above, he was considered a true moral philosopher, and a proponent of nonviolent resistance, which had a profound influence on many world leaders, among them the great *Mahatma Gandhi* and the American civil rights leader & Nobel Peace prize winner Rev. *Martin Luther King*. The most recent *statesman and freedom fighter* who became a Nobel Prize laureate was Nelson Mandela of South Africa.

"Well, now, this music is an elegant waltz, played in grand European wedding receptions for the bride & groom and the guests to waltz through the richly bedecked ballroom, under the delectably shining and sparkling crystal chandeliers dangling from the gilded ceilings.

My parody for it, is titled "*It's a beautiful day....It's my Wedding day!*", describing the dreams and feelings of a soon '*bride to be*' on her wedding day".

Elise is flabbergasted! She feels that she is in an exhilarating romantic dream; experiencing fairytale happenings; a fantasy world which is factually correct in all aspects, and respects.
She is wondering how all these situations and themes are coming alive in such closely connected and mysterious ways as this.
"Is it a plot set up by the TV network?...Am I in a "*Factual day-dream*?"...Is it all real?...Am I feeling romantic hallucinations?...
Do these happenings manifest in its natural, random course of events and occurrences?" She felt confounded by many divergent thoughts and plausible probabilities!

Whatever they may be, she decides to enjoy the moment to the fullest! So far so good! "Let it be.. let it be...!; a rare sweet moment in life; Worth relishing the whole lifetime".
"*Oh, my God!*; let this moment shall not end...!...*oh, my Lord!*...please!...help me!...please...!

She seems to have lost her '*stage presence*' and was overheard frequently exclaiming in soliloquy (a monologue, talking to oneself). Inadvertently, it gets through her hidden microphone fitted on her for singing. The sound engineer and the director let it go on air, raising consternation and dismay among the audience. All are mystified!

Not knowing the significance of any of it, the audience become bemused by the drama, Elise's spontaneous reactions, and obvious consternation.

Her emotional commotion is self-evident. Justin & Gemini are summoned to her side for the much-needed support. They stood in close proximity watching her in great alert, and attention. The cameras did not miss a single move, moment, expression, or reaction. The video director at her console is now presented with too many images to show and share with the viewers. She struggled to find out ways. So, she asked her assistants to compile multi-view montage (a single pictorial composition made by juxtaposing many pictures) consisting of split screens for a full-screen-display, and a different one for a crawl that moves from left of screen in continuous loop. The crawl shows current camera shots with constantly changing views from the stage and the audience. In order to make a multi image full screen collage, each camera image is proportionally *scaled* to smaller size and placed in each quadrant of the full screen. In crawls, the individual images are run in a sequence one after the other. Both these effects show the simultaneously occurring scenes.

Split screen multi-image-montage shown on TV screen

Usually, the audience sits in the dark; have only the dim '*downlights*' to barely see the immediate surroundings for safety reasons. But now, all the lights are turned on, shining bright, to help the cameras catch the individual audience-reactions. The cameras are all UHD (ultra High Definition) too. They show faces lot better than one can see himself / herself in the mirror, not missing any spot, or mole; wrinkle, nor dimple; frame, geometry, or countenance (the expression of the face).

The audience is actively interactive, unlike the usual habit of sitting still in their seats and turning the ears to the music. The local population is by habit and tradition, not very revealing or expressive in public. There is absolutely no expression of PDA (public display of affection).
Because!
The '*self-styled*', 'self authorized' militants, or the '*moral police*', the local equivalent of '*Muthawa*; will be quick to pounce; take severe, punitive actions if any violation of the tradition is spotted!

At this instant and place, the entire ambience and the Television production has been transformed to its highest potential and effect. As they say: "*necessity is the mother of invention*". It is clearly in evidence here!

The applause subsides; the audience takes their seats. They seemed to like the theme. Most of them are with family members; some for the first time out for a good time alone as couples, away from the nosy parents, and in-laws. Present is a number of newlyweds on their honeymoon; some secret lovers and dating couples hiding among them. Following the televised broadcast, some revelations will surely happen; some may be caught red-handed; some may be subjected to stern questioning and reprimands. There will be video footage evidences of individual's indiscretions (lack of discretion; imprudence) as proof. Damn! There's no place one can possibly hide these days, thanks to the Television, surveillance, mobile phone cameras, and GPS location devices!
Heck with it!
For them, for now, it's a moment to behold! a sweet, tender, intimate, and enormously gratifying one which may never come by again. They ascertain themselves "let's continue to enjoy, whatever may happen, may happen; all that'll be later;...We'll deal with it, then and there. Let nothing upset our good mood!"

The Piano sets the rhythm for the slow flowing elegant dance. The accordionist switched his Register to *musette*; picked up the tempo and the key in D minor; starts with prolonged *euphony* notes of A_3 & A_4 and a beautiful slow sliding glissando downwards collapsing in the lower key of A_3. The violin starts the melody with nice smooth "*full bow*" action; the bow nicely balanced with the same amount of pressure, up and then down to the end of the bow, playing a dainty legato (smooth, blending notes). By doing so, she is able to play the notes continuously without any breaks in notes. She pretends she is also dancing and twirling to the music in the ballroom; Enjoys herself!

Her mother whispered to the stage manager suggesting dimming the light on her violinist daughter. Suddenly, her dangling pair of earrings are seen scattering bright beams of colorful light as if they were studded with expensive diamond jewels. They were merely some cheap decorations, but tastefully selected crystal colored '*rhinestones*' (sparkling artificial gems of diamond imitations made out of tiny luminescent molecules and pigments of RBY of the primary colors- Red, Blue, Yellow from which other secondary colors Orange, Green and Purple can be made by mixing two primary colors). They were selected and hand-made by her father. He mounted them precisely in certain geometric shapes and patterns using theories of light reflection, refraction, and absorption, studied in Physics classes many long years ago. They are encrusted and inlayed in the shape of a fully blossomed, bright lotus flower. It was made from articles and items secretly taken from his wife's collection of fancy knitting stuff. It was a three-month-long project undertaken to demonstrate his paternal love for his '*princess*'!

Pigments are molecules that give color by effectively trapping light. Pigments of different colors absorb different wavelengths and hence have its own color, tint, and brilliance. Chlorophylls trap blue and red light; shine as green. Carotenoids absorb bluish-green light; appear as the '*carrot*' colors- red, orange, or yellow. Pigments in the plant molecules are associated with proteins which are light sensors that attract them to the sunlight, also, to be attracted to one another for the purposes of photosynthesis and multiplication. This is one of the infinite numbers of wonders of the enormously resourceful nature!

The luminescent materials transform invisible (infrared and ultraviolet) forms of energy into visible light! In other words, they emit light under external energy excitation; reflect incident energy in the form of electromagnetic radiation some of which fall in the humanly visible spectrum. Her father has also made use of minute prisms in different shapes that break up and scatter light into more intense and uniquely beautiful spectrum of colors. Those days, he was regularly seen wearing a '*watchmaker's magnifying glass*' while nimbly assembling his modest masterpiece.

Her jealous and wealthy friends used to ridicule and bully her saying "*It's all fools gold, honey*!" She endured the bullies and insults; but boldly retorted "it's a gift from my poor daddy, lovingly and painstakingly created for me and given to me on my HS graduation, to show his appreciation for my hard work and achievement. It's more precious than any diamonds cut in Antwerp or Bangkok. Besides, mine sparkle brighter and exude many colors, not just a single boring drab color; mine jingles; yours just a dull *thud* (a dull sound without resonance), or none at all. I don't need any other,... none any more!"

With her playing style, the ear rings dangle, glisten and twinkle with sparkling luster. The incident light on them disperses and reflects various colors. Like diamonds, it shows a very high degree of transparency, reflectivity, *refractivity*, (the physical property for bending or deflecting light) and dispersion which results in a '*fiery brilliance*' and a display of prismatic colors.

The program director on her Preview Monitor, detects the intent and curious stares of the audience at the dazzling earrings. She asks the lighting technician to momentarily *dim* the stage lights down and hold it for few seconds and then slowly bring them back up again. She further instructs him to cast a narrow width spotlight beam on to the earrings for the same period.

Promptly, he complied. Momentarily, this produced the desired bright radiant '*comet tailing*' (a long, bright blazing tail) effect. For the duration of 7

seconds, the iridescent (displaying a spectrum of colors that shimmer differently at different angles of view) earrings were the only things shining on the stage. They were seen like twin brilliant stars shining in the distant galaxy!
Vow!...The Miss. Violinist looked "*radiantly musical & beautiful* !"

When the light came back, the audience whistled in amusement, appreciating the special effect. The cameras captured all the transitions in its real time, beauty, and spectacular effect with tight close-up shots.

On the stage, when anyone plays a solo, the other two soloists look on appreciatively and with ostensible pride. Each nods with a smile in earnest approval, and admiration; even swaying their body in tune with the musical movements. Giving appreciation to others then and there is a rare trait, and gesture in the society using the pretext that '*people may feel hubris in their accomplishments; that's a sin*'!

Now, the accordionist switches Register to '*Cello*' (Frequency Range: 65 Hz - 1.0 kHz) for the second refrain, as it is the most suited for the passage. It sounds somber and serious!
The union of two people in matrimony is indeed a solemn affair to behold and cherish for life!

At the end of the refrain, comes a passage, where a set of Pizzicato (Played by plucking rather than bowing the strings) has to be played. She plays them with her fingers plucking the *A* string one after the other, (finger staccato) then a sudden accent with a forceful down-bow motion. The down bowing is preferred for long and heavy note, utilizing the heaviest and tightest part of the bow, that is near the handle.

The Piano and accordion join in, to embellish the climatic moment with their '*fill-in*' phrases!

The audience loved it; few jerked themselves off their seats spontaneously and in sympathy; then relaxed back on to their seats as the music changed its tempo in to a more relaxed normal waltz tempo. It is the most opportune moment for the male dancers to swirl their partners in a wide circle around them. At normal times, they stride one way; finish the stride on a tiptoe leaning and extending in the direction of motion, as if they want to catch something that is beyond reach; reaching the limit, regain balance; gracefully stride back in retreat in the opposite direction.

A post on media from one of Elise's cousin Alice, reads:
"Listen folks! Today, 31st August, is Elise's wedding anniversary!...

The next post shows Elise's mother's response:

"Yes, you still remember, Alice?... you were one of the flower girls that day., you're still cute as you were in those days!...". The date gets an instant confirmation.

"Elise has now been married for three years ....no children!" one person exclaims

".....not yet?"; Sujatha, a onetime classmate wonders.

"......it could happen anytime; perhaps tonight!"
an anonymous, made his/her naughty comment.

".....She is an alumna (Latin: a female graduate or former student of a school, college) of KIS (*KodaiKanal International Residential school*) studied IB (*International Baccalaureate*), French & vocal music ..." Sulochana Guansekharan, Kodai

"....Kodai is mainly a tourist hill resort at 2200 m height, in Palani Hills, T.N..."

"......Oh!, that's where she learned fluent speaking style. Yes, sure I know. It's an English medium school!"

"......Also, my son Gemini is an *Alumnus* (male graduate) of the same school...took IB, German & Piano lessons, passed 2 levels of piano playing proficiency tests conducted by the '*London School of piano music', He joined the band with his sister. They are very close to each other*."
The proud mother Celine tweets hastening to give some proud recognitions to her son.

"....Elise wanted to learn swimming; took lessons at the water sports club by the '*star lake*' behind the school complex..." Celine added.

"... interesting!; must be from a rich, aristocratic family! Good education, training, and upbringing make a big difference which I'm not blessed with. I envy these folks. Quite frankly. I'm a communist comrade!"
M.V. Dasan Srinivasan

"...the lake is shaped like a star, man made, from collecting waters flowing down the slopes from the surrounding hills; no waves, you can row boats.... I had my honeymoon there...we rowed a boat around the lake; very nice, enjoyed it very much!" *Dr. Naresh Kumar & Gayathri Devi, Boston*

Well, the music goes on to another refrain (recurring passage) and reaches the highest note D5 : reaching towards the upper end of the accordion keyboard. The violinist slides her fingers closest towards the violin *bridge*, reaching over into its body of the violin to play the same note, an octave higher. It was carried out in a graceful *finger slide*.

The brass group in the band shows great excitement. The saxophonist jumps up. Flamboyantly, he started playing the variation of the refrain in *impromptu* (prompted by the occasion) and facile skill. The audience reacted with long applause. That prompted the trumpeter to follow suit, who made his rendition exceptional with his full breath blowing. Not to be outdone, the trombonist played his version masterfully moving the '*slider*' in smooth style. Meanwhile, Justin, invites the trio to join-in. Together, they repeat the part the trombone just completed. The listeners are overjoyed by the beautiful interlude and spontaneous interactions among the musicians.
As a matter of fact, one of the refrain in the original composition was specifically done for the brass, precisely the trombone. Now the music feels full and complete like the original masterpiece itself!

The music comes to an end in a dramatic high note in "**Pa <u>Pum</u>**" executed in *Crescendo & Sforzando* (strong accent & attack) with a heavy down-bow by the violin; a D+D^m bass chord on the accordion played simultaneously. The piano did the same notes in superb synchrony with an abrupt stop at the final note. It was a sensational ending, enjoyed by all.

The captivating ending captures the imagination of the entire audience. They stand up in salute showing recognition for the excellent *arrangement* (an adaptation of a piece of music for performance in a different way) of the music they heard, and the delightful sequences that occurred. This is the first time, the local audience is exposed to such a dramatic and heavily articulated rendition. It goes without saying the response was unparalleled, unprecedented. Elise, and the entire band jumped up along with the audience; the auditorium erupted in clasps of Waltz rhythm. Some have already learned the waltz rhythm pattern- counting **1**, 2, 3 accenting 1; 2 with a weak count; and 3 with no accent at all, thanks to the Pianist who played it wonderfully, as a prominent part of the music in '*ostinato*' (a '*chord progression*' that is repeated over and over as a musical accompaniment).

The traditional singers, musicians of Bollywood & Mollywood stars occupying the front rows are astounded at the audience's astute (having insight or acumen) appreciation of the Western music they heard and that they enjoyed it in great measure. They feared that their own standing and reputation may tarnish; domination diminish if they fail to improvise new, special presentation techniques, and include some showbiz antics. They cannot continue singing the love-songs from movies of yester years for their entire career span. They must innovate, fuse genres, garner attention, encourage participation, and enhance their music by adding more instruments to both rhythm and melody sections. Piano may be the one, as it can play both rhythm and melody; ah,...'*two birds in one shot*!' They mused and made plans for the future. It needs work; a lot of it!

"There shall not be any hesitation to adopt better things, irrespective of its origin, ethnicity, or style. When something is proven good, it must at least be considered. People accept foreign goods, technology, way of life, customs & practices, nuances, languages, sports, and even religions. Why not music?. We consume, join, learn, and practice things invented and introduced to us. We adopt and adapt by its attraction, usage, for a purpose, to achieve, to profit and/or benefit; by intent or force, ultimately to enhance our lives. The world is a dynamic, interactive, mutually stimulating melting pot, more aptly, a boiling cauldron (a kettle or vat, used for boiling).

We reside and live within it; become integral part of it, voluntarily or otherwise! These are plain *facts of life* and true *state of affairs!"*

"There will be challenges along the way". They mused; reflected.
It's one of the many *lessons* from this evening's experience.

Astonishingly or not, the audience are still insatiable; their musical acumen have blossomed and peaked. A joint unrelenting shout echoes throughout the hall. The sound level display signifies that it is the loudest intensity in *dB* (Decibels, unit measure of sound level or loudness) of sound and by far, also the longest duration recorded. Records are set and being broken many times even on the opening day; just incredible! The camera switches to the display monitor inviting attention and notice. It only helped to breach the barrier and establish yet another record eclipsing the previous one just set. The tech-savvy audience, switched on their Apps on the smartphones, and '*fact-checked*' for an instant affirmation. They recorded it, saved it with a date stamp, adding their own individual '*voice tag*' to it. This is another testament that they did attend the opening concert '*in person*'!
Someone in the crowd remarked "*only a sports arena can be as loud as this*!"

The audience wants more! Thy are exhilarated! They are possessed, attained a trance (a dazed or unconscious condition) under the influence and power of music. Their loud demand is unrelenting and uncompromising!
Ostensibly, the family singers are getting alarmed and frantically checking time again and again; staring at the watch and asking confirmation from others. They glance at the curtain; hopes the embroidered and frilled brown curtain drops down at this very moment, so that they can run away from here. But, nothing happens. The lights and cameras are on them. They look stunned, embarrassed, and unsure, as what to do now... They have things to do, rather important things. Those are not under their control at all!

As a matter of fact, it's quite urgent; peremptory! (imperative; precluding or not subject to debate or question). But no one else knows; themselves alone! That's all their own concern, and the only concern they have at the moment!

But, they find no way out other than somehow appease the crowd, and do it rather sooner. That's possible only by offering to play another one, and do it in a nice and affable manner; quite unpretentiously, showing no undue haste or irritation!

"Well, the next one has to be the very last. Elise is waiting to perform her own show, and many are eagerly waiting to hear her superb voice. We happened to be mere intruders and disruptors. Sincerely, we apologize to all!
But, before we begin the last song, we want your solemn, inviolable promise!"

"We, the people, solemnly promise!... So help us God!...
For sure...! Definitely..! Definitely, positively...!.. Yes, the last one,...
of Course! Indeed...OK...Now, C' Mon!..,
Let's do it...Hey folks, Let's behave!...Cooperate,...Be good!"

The hall resonated in various assurances and promises; echoing from the side walls and the dome. It seems that everybody agreed and answered in good faith.

The producer announced from the stage "Ladies and gentlemen, our esteemed and wonderful music enthusiasts, please take your seats and let's hear the last music piece from our honored and the very enchanting guests of the evening; it's going to be a fantastic one too...., I promise you that!"
"We'll continue the '*First Concert*' with Elise soon afterwards!"

"This is one of the most beautiful waltzes we know how to play. I have lovingly written my Malayalam Lyrics for it, to help you sing the tune in our mother tongue. It's called: " *The perennial Pampa flows forever..!* ".

This original music was composed by Josif Ivanovici, a Romanian composer. He called it the "*Danube waves*".

In the interest of winning the crowd and showing no undue haste, the accordionist gave a narrative '*intro*' to the music, as he has been doing throughout this course of adventure.

"The Danube river is the second longest in Europe, second only to the mighty Volga. It flows from the Black Forests in the Baden-Württemberg state in Southwestern Germany from the slopes of the Alps to the Black Sea, through many countries. Many picturesque cities have sprung up along its banks, Vienna, Bratislava, Budapest, and Belgrade, to name a few (they are capital cities too!). There are resorts, wildlife, and bird sanctuaries built on the river sides. It's the home of the biggest freshwater fish the "*Beluga sturgeon*". The costly '*beluga caviar*' of the Black Sea & Caspian Sea basins are the ultimate taste of the rich, famous and connoisseurs of the world's fine cuisine. Only they can tell us the taste of it! The cruise through the river is one of the most magnificent experiences for tourists. There are many songs and masterpieces about this natural beauty, capturing the imagination of millions of population living on its banks and many visitors who come to see and enjoy.

I could very well relate it to the beautiful river Pampa as I've many times traveled up and down while visiting my mother's family on every school vacation. The magnificent river is lined with palm trees and coconut groves; paddy fields and haystacks; bungalows & buildings; churches, temples, and schools. It originates somewhere in the Sahyadiri hills in the *Ghats* and flows down westwards around most of the *Kuttanad* basin in Central Kerala, irrigating, fertilizing, making farming, rice, coconut cultivation, as well as water transport possible. Finally, the river rests peacefully in the vast embrace of Vembanad lake. The natural beauty and contributions provided by the river are manyfold and invaluable to the citizens of the land!
Furthermore, it is especially dear to farmers: a lifeblood of my own family fortunes and indeed to myself, personally!

We thoroughly hope you will enjoy the song tremendously!
It is otherwise popularly known as the '*Anniversary Waltz*.'

Just close your eyes and imagine how the Pampa river flows across the landscape, may be even in front of your village or house! The music will help you, in the process, to enhance your imagination as well as building respect for all the given natural & environmental resources we inherit from mother nature."

Elise thinks to herself watching Justin: "Today, Justin & I need no other anniversary celebrations of our own. We could never outdo this one! Honestly, we could not do or wish for anything better. We must do something wonderful, which could be donating money for some noble causes. First I've to ask Justin! Just look at the audience!... All are rejoicing; are invited to our '*Anniversary celebrations*'...This family makes our memories so remarkable, unforgettable, and enduring!...May God bless them!...How wonderful it is...it's a blessing...truly a '*God-sent*'. I wish I could talk to Justin at this moment and express my true feelings. Justin and I should live up to it and become exemplary couple all our lives! That must be our solemn pledge in this very moment!!

The overture (prelude, an instrumental composition intended as an introduction) begins with the piano playing in energetic intro "Pum, **Pa**, Paa; Pa-ra, Pa, Pa, Paa"; a few times followed by a hasty '*flourish*' of flowing galore of ascending notes building up tension (swelling water pressure); then descending towards the tonic note (releasing the pressure into a stream); and finally ending the phrase in delicate tones of Pum, pa**pa**, Pum, Paa... The accordion repeats the exact passage in the same manner reminding the listeners once again of the charismatic persona and relaxed character of the fabulous river. A hesitant pause follows indicating an obstacle in its path. The water collects itself in a deep puddle before flowing over it and freely soon afterwards. The violin leads the music in flowing manner by playing the '*artificial harmonics*' using two fingers, the 1st & 4th on the same string. (Please Note: the violin is played with 4 left-hand fingers, leaving the '*valley*' of the thumb to hold the instrument. The thumb is not counted as the 1st finger, the next one assumes the number 1 spot!). The resulting sound and the style provide a feeling of gentle, smooth flow of a stream and the breeze gliding over the waves in continuous embrace in its long journey that has only just begun. It replicates the very same manner the river Danube forms its untraceable origin in Schwarzwald (Black forests) in the Alpine valleys. It is the collection of water that melted from the snow accumulated over Winter, true to the description in the song '*My favorite things*' in *Sound of Music*: "*silver white winters that melt into spring*".

Along the way, many other streams coming from many directions deposit their share of water into it joining forces with the major river. They are also initially formed by drops & trickles of water, rolling down the slopes by gravity. They continue flowing along its natural '*path of the least resistance*' forming canals. Several canals make a bigger body of water, the river. The river swells, surges, picks up speed, and relentlessly flow forward. They form tributaries going towards different directions along its run; makes its way forcefully forward; tenaciously goes over, or around obstacles in the path; finally, falls into the vast seas, the place of its final destination, resting place, and salvation.

The accordionist rolls his chair, sitting on it, scoots over to his daughter. He switches the Register to '*tremolo*' (violin sound) and joins her to play the melody. The violin has its own natural tremolo sound. The two tones blend harmoniously together, producing a wavy ripple effects, most suited for the theme and mood of the music. The Piano provides the rhythm, and plays '*fill-ins*' and accompaniment, suggesting the meandering paths of the river and sudden changes in its slopes, the laborious climbs on to many obstacles in the path, and speedily gliding down the valleys. The piano articulates the scene by playing successive staccatos in ascending or descending order, indicating the amount of stress as well as the direction of the flow.

Soon after its weak initial formation in the Schwartzwalds, and after flowing a short distance, mysteriously, the stream vanishes; perhaps still flowing as an invisible underground current. In fact, there are constant clandestine tussles fought out between the *Rhine* and the *Danube* rivers for the share of the precious water. The Rhine flows through Germany, along the Rhineland, border between France &

Germany; through short stretches of Austria, Switzerland at the origin, and on to the Netherlands, eventually emptying into the North Sea while Danube flows Southeastwards towards the Black Sea.

The water can't ever stay still; it has to move; continuously, it flows. It may not be always visible. But there are tides and ebbs, undercurrents and visible currents, whirlpools, surges, and floods, also bubbles, and waves. The body of water reflects light, the moisture in the atmosphere split the light into spectrum; sometimes so visibly and spectacularly projected onto the vast horizon. We call it the *rainbow*. Contrary to our notion that the rainbow is an arch, it is actually part of a huge full circle, better seen and understood from much greater elevations. A photographic image recently taken from the 460 M Russian skyscraper, and another from the mountain-tops at Yosemite confirm this fact. The earth's curvature, provides a distorted visual illusion as an arch, since only the upper half of rainbow can be seen from the ground level with a naked eye!

The setting sun seems to say goodbye with a '*good night*' kiss of the waters in its daily parting of farewell. The many exotically colored shapes, patterns, and light reflections seen in the evening sky mark the '*joining of forces*' by the nature to reassure the living beings that they shall return, on time to wake us all up in the next morning.

The composer has melodically elucidated the scenes and phenomena with his musical expressions and phrases, carefully selecting notes and tempo. This is now being articulated by the trio of instrumentalists in clearly perceptible, and delectable ways, reminding us of the wonders and beauty openly visible and felt around us.

The Danube meanders along as many as 10 different countries, (Germany, Austria, Slovakia, Hungary, Croatia, Serbia, Bulgaria, Romania, Moldova, Ukraine) and flows onwards from the Romanian delta to the Black Sea. It, being the Europe's only such major river, has gained special love and significance. The river inspired the imagination of many artists and musicians. It helped the cities and countries in its path to become centers of arts, music, culture; and power. The audience is now capturing a piece of the same spirit and enjoyment. The famous Waltz by Johann Strauß II "*An der schönen blauen Donau*", (On the banks of the beautiful blue Danube) immortalized its glory and praise among music and nature lovers.

Flowing from West to East, from Central to Eastern Europe, the river has now met a turbulent patch in its perennial journey, where the banks are lined with gray, grim-looking granite banks and sharp ridges where the water becomes wildly rough, turbulent, and difficult to tame, and navigate. Up until the recent past, steam powered locomotives were used to pull the cargo vessels using iron/ steel chains further upstream to the tranquil waters of the ports. This particular *stretch* is heavily infested with snake and other venomous reptiles. Now, a dam built upstream from the spot, supplies Hydroelectric power and irrigation water to Romania, Serbia, and other Balkan countries.

It becomes all downhill and downstream from the *Iron Gate* and past the *Carpathian mountain range*. The river flows freely the rest of the way to the Black Sea over the long, sloping, pastures and swamps of the rural Romanian, Moldavian plains, and the shipping ports of Ukraine.

From the music, listeners can discern a feeling that, as the river finally reaches its long (2900 km) traveled destination; becomes so happy, buoyant, and excited; gather kinetic (caused by motion) energy, scrambles and splits into multiple rushing streams in many directions. In its frantic hurry, it creates deltas, estuaries, lagoons, sandbars, and coral reefs; then peacefully embraces and rests in the vastness of the Black Sea, mixing its fresh water with the salty seawater. The abrupt confluence results in some jostling, agitation and turbulence. The music ends reflecting the same proceedings with its symbolic tunes, and phrases.

The captivating occurrence is reminiscent of the everyday scenes from the Arcadian lifestyle when a toddler who has been waiting for its mom to come home from the remote fields; upon seeing her, it runs weeping in joyful sobs towards her, breaking all restraints. The mom who has been equally anxious takes her baby in loving embrace who suddenly lulls (quiet down; a sense of safety) as it rests on her bosom, calm, warm, and in sublime bliss!

In similar fashion ends the music, strong and true to the character of *The Danube River* when meeting the sea after its arduous journey started from far away mountain valley.

The delta and the region so formed serve as favorite '*transitory and nesting place*' for migratory birds prior to Winter, before they proceed to fly onto the warmer climates, possibly to the continent of Africa.

The audience slowly rise from their seats, knowing that the beautiful concert and the supreme sounds of the evening are sadly and inevitably coming to the end. They could very well relate to the song as they have seen and often traveled along the Pampa many times before; had seen familiar sights it offers along the way. They are happy that they were able to experience something unusual which will help them in many ways- the appreciation of music, the happiness, the stories they heard; the events that transpired, new visual and audio techniques that enhanced their overall indulgence; the thoughts & feelings it brought out; the tenderness & loving feelings people feel towards one another; a new awareness of the beauty and the marvels of our environment which we usually take for granted; the need and urgency to protect and maintain them in a way we must, for our own prosperity, and the survival of the only living planet, our own blue planet. The rivers are not only the life-blood of nature & environment, but also of the industrial life in moving goods to the consumers and materials to factories and power plants. We must guard and protect them in responsible ways. No wonder, Scientists are looking for traces of water on Mars before determining if it is livable, viable, and friendly planet.

The audience feel that they are greatly enlightened; find themselves in fine mettle. They are now better able to respond to their spouses, children, parents, siblings, and fellow-beings in more affectionate and respectful ways. It will certainly become an essential refinement in their own conduct and family interactions. Everyone has feelings and desire to be happy. There is more to life than work, business, money, status, self-importance, ego, and pride. They feel confident of the fact that they have learned something, actually, a good many things; enriched and edified themselves through this concert experiences. All in all it has been a strongly impactful evening in their lives. They are determined to '*make a difference*' as citizens and in their respective roles! Further, they have a good story to tell among loved ones, relatives, friends, and colleagues at work places. They felt wholesome; had a new awakening of sort; an illumination of minds and a reinvigoration of certain dormant faculties!

They are clapping as they entertain these thoughts deep in their minds, respectfully expressing their appreciation and gratitude in a way an applause can best carry such sentiments.

The curtain falls for the intermission. Everyone takes a short break; looks around; see only pleasant smiles of joy and satisfaction on faces.

The show resumes shortly; so as to make up for the extra time lost in the '*unscheduled*' events that took place.

Elise took the mic and sang several songs premiering her new album. She entertained the crowd to a range of musical styles, from well-loved film classics to folklore. Before the start of each song, her manager gives a brief account of the theme, the *muse* (guiding inspiration) and a summary of facts and information about the song as radio announcers do before playing the music, such as the key, raga, rhythm, tempo, mood of the song, etc. The orchestra complemented her skills with innovative articulations, ornamentations, and delicate sound effects. They used new techniques, harmony, polyphony, phrases, and voice effects. Each song is embellished by a pleasing solo instrument while the others expand the secondary parts. There are several improvisatory interludes in which different categories of instrument take the lead and turn; dictate rhythm and phrases; then hand it around to the others in the band. The orchestra and the audience enjoy these episodes of interactive styles. The performers gauge the liking and interaction of the audience and probe the extent up to which they may go. Sometimes, the audience takes the lead and prompt them to a different direction by hand eye manipulation and subtle gestures. This is a very interesting and integral part of the traditional Indian music '*kacheri*' (concerts). This practice transcends modern genres and styles.

The concert finally comes to an end with each individual instrumentalist showing off his/her newly '*arranged*' skill-sets. After one or more rounds of such ostensible display of technique and dash, they join together as a band playing euphoric, and exultant potpourri of sounds and music bringing a befitting culmination to the glorious concert, '*The first concert*'.

The audience is extremely pleased. They clap and show their unbridled appreciation to the artists. A few Bollywood stars climb up the stage; congratulate members of the band. The female singers warmly embrace Elise. The Bollywood singers used to think they are the *ultimate* in singing. Their hubris is belied here in this concert. It's a new, surprising, and revealing revelation.

The musicians profusely greet band masters, and their favorite instrumentalists; ask them to show how a particular technique was renditioned. It's all cheer, joy, and camaraderie. Some of them take a good view from the stage; survey the audience galleries; make a thorough mental assessment of the concert hall; take pictures, and video. They ask the audience to pose for them; talk to them cordially, building rapport or perhaps doing some PR work. People immediately recognize the film stars and performers in their own right and fame. Many of the singers are described as "*playback singers*" as they are seldom seen; they sing mainly in the recording studios. They are recognized by names and not by their faces! Having attended this gala, they are excited to perform live; foresee some potential to do it at this very venue sometime in their career. Soon, they'll be calling their agents to explore possibilities.

In all respects, the evening has been an enchanting one. People cannot wait to get home; turn on the playback devise; do an exciting personal running commentary of what *I was seeing, thinking, feeling,* etc. to all the rest of the family members who were left behind at home.

In a concert, according to conventional norms, the '*finale*' is always the concluding part of a musical composition. However, here today, things don't follow customary conducts or standards as one would normally expect.

Every show must end!
It must end in a certain way, with pleasantries, formalities, curtsies; goodwill and good words as the glorious circumstances warrant. Many people have worked on and off the view, on- stage and backstage; made all the preparations; planned, toiled with a sense of pride and perfection to bring the show to a pleasant conclusion.

Preceding the farewell message by the lead singer, the Program director made her rare appearance, thanked the audience for their participation and support; being appreciative throughout the evening, all along the twists, turns, heightened emotions, and elated ecstasies during the eventful concert; their passion and enthusiasm for the performance. She thanked the Sheikh and all the other magnanimous benefactors by their names for their overflowing generosity in building the picturesque complex and the exquisite concert hall equipping it with the most advanced technology in each of the several categories. It shall stand as an exceptional artistic enterprise! With those gracious words of praise and gratitude, she yielded the podium to Elise.

Elise still in a *trance* (a state of ecstasy or mystic absorption so intense as to cause a temporary loss of the presence of mind) and exhilaration with joy and emotional turmoil, began in an unusually unsteady and stuttering voice:
In the foremost, I feel so very fortunate to be invited to perform in this grandiose, imposing, and splendid concert hall in front of this brilliant, exceptionally musical, and appreciative audience.

"This has been a special, memorable evening for all of us. We discovered something together which enriched our minds, memories, feelings, and thoughts. It was especially meaningful for me, personally. I'm inclined to believe I found somebody I have been searching for.

It is a moment of discovery and redemption!

I did also find humanity, charm, gifts, talents vested in others; in every one of us in various ways and different depths. It humbles me, it gives us new understanding and vision; joy in ourselves, and other people. It reminds us of the belief that we, as humans, can achieve greater things together, accepting one another, working together, harnessing our talents, and capabilities for the common good for our own destiny, and a better future for the posterity. This grandiose structure, the fabulous complex on which it is built, the magnificent occurrences happened within these resounding walls, as well as the parade, procession, and ceremonies that are held today, confirm our beliefs, and convictions.
I leave, just like you, enlightened, discovering humanity, goodness, and power in ourselves. I am moved,... tremendously!
Frankly, I detect the same in you too!

"Ja...Yes, indeed!....of course....you're perfectly right, Elise!"
The audience responded in confident reassurance.

I had many thoughts. During the concert, with many things happening around at a fast pace, and succession, mostly in mysterious ways, I had to steal some time to discuss few important ideas with my manager, The TV executives, and the band. We did come to a unanimous decision to donate half of all proceeds to the music institutions, academies, and the organizations promoting and contributing to the human spirits in certain positive ways.

Nothing can take place without the patronage, participation and the enormous support from the many ordinary and skilled people. We perform for you, the audience, assembled here, those who are watching at home on Television, and those who might hear about it afterwards through replays and playback of recordings.

You provide us a platform, an opportunity and a challenging desire within us to entertain you the best way we could.
It is music for us!...really! Perhaps, the only thing we can do well in life to some degree of proficiency. At the same time, we do recognize and appreciate the fact that you are all very good at what you are doing, in your pursuits and careers!

You lent us your ears and coveted presence, sung along; you felt wholesome and appeared jolly. That's quite satisfying to us. The reactions, visual feedbacks and the vibes emanating from you, invigorated and excited us. We played and sung our hearts out on top of our sounds, in exultation. We derived reward and satisfaction that we could entertain you; at least momentarily helping you forget woes, and worries. We, all have plenty of them. Don't we not?

Now, about our dear band!
We are a close-knit group of friends, neighbors, classmates, and relatives forming a cohesive band in goodwill, harmony, and melody. We have great understanding and respect for each other and their respective abilities. We are obviously bound by our passion for music. Each has an aptitude or special talent with the chosen instrument; capable of listening, and adapting to the rest. The orchestra is part and parcel of the music. It provides the essence, inspiration, background for music; create mood, character, set the singer for the music you could remember in words and helps you sing for yourself. It is unlike the beautiful classical music compositions devoid of lyrics or words that a person can sing with, or '*sing along*'. On the other hand, the songs you hear are universal enough for people to enjoy; they clearly state the themes, create proper mood, and provoke sentiments and thoughts. The band complements the music in many untold ways with a '*repertoire*' (the skills, techniques, etc., used in a particular field or occupation) of embellishments, creativity, ornamentations & articulations, fill-ins, rhythms, effects, flourish, and beauty. They help to package everything in a pleasing envelop of rich sounds, resonance, reverberation, tempo, style, and fanfare; elevates the listening pleasure to an exalted level! Unrestrained, you expressed your exuberance too.

That was phenomenal! We are greatly gratified.

Then, slowly, Elise begins to introduce each member of the orchestra.

While she calls out a name, the performer briefly shows off what kind of noise he can produce; then comes forward in great humility and elation. He or she takes a courteous bow with folded hands. The audience applaud the performer's skills until they go back and take their respective places. Most of them were familiar faces to the public. The process took some time to finish in its entirety.

The audience has been waiting with restrained reticence and patient curiosity to see and greet the '*family musicians*'. They were totally unknown and unseen before; nor heard of. The atmosphere they created had been a pleasant surprise to everyone though no one knows anything about them. The proper time has arrived!

That's now!...
yes, this is the moment everyone was waiting for!
With great excitement and expectations, Elise looked joyfully and expectantly towards the program producer to get the list of names handed over to her, as the President wants to be supplied with a set of "*flashcards*" with '*keywords*' to ceremoniously say or utter in a pompous speech. It must contain an introductory description of each, their names, qualifications, and other glorified things about them.
All that is required of Elise, is to simply read out of that paper or from a *teleprompter* screen in her personal enthusiastic style with occasional embellishments and relevant attributions!

"Now, is a sublime moment for me" she began ebulliently (in high spirit). Then suddenly began to choke for words. Obviously, she is struggling to control her runaway emotions; paused a bit...wiped off some tears of joy with her shapely trembling fingertips'...slowly recovered to a semblance of balance, poise, and composure.

In a shivering voice, she spoke: "......it's quite an eventful evening of new experience... we traveled through unchartered territories...felt like we have traveled far and wide...; by rivers... danced in the ballroom....felt passions & pangs of love, and nostalgia...visited idyllic scenes, places,...came to know great masters, brilliant minds, and the legacy they left for us...Our vision and world have widened in the swiftly passing moments...The scope and possibilities for us now exceed our senses; still abounding (plentifully supplied)!
Now, let me relish this moment of honor and privilege,... of introducing the magnificent trio of '*family musicians*' who stole the moment and our hearts with their affable curtsy, and heroic repertoire."

The themes and music titles they played

It's all silence!,.... a void silence hangs in thc air!.....
It lingers and spreads everywhere in the hall. Nothing takes place; not even the usual hustle, bustle, or rush towards the exits. No movement is seen on the stage either; only blank, stray looks in various directions, reminiscent of the scene from the movie classic, "*Sound of Music*", when the award was announced to the patriotic '*Von Trapp family singers*' in the *Salzburger Musik Festspiele* (Salzburg music festival). In the fact based story, the Von Trapp family singers had fled soon after their stage performance to evade the Nazi Gestapo agents who were skulking to forcibly take the former Austro-Hungarian Navy captain, Georg Ludwig Ritter von Trapp to Bremerhaven to fight in the Nazi Naval force, soon after the

award ceremony. Here too, no one was there to appear; whether they fled or not; no one knows!
Elise blushed, showing some exasperation.
Yet, she waited, patiently.

Ms. Bhavani Bhagavathi, the program producer came forward; once again introduced herself and said:

"The family had to depart in a haste to the airport to catch their flight overseas. They had checked- in earlier online and had maroon colored boarding passes with them, indicating it were possibly a *Qatar Airways* flight. Hurriedly, they packed their accordion back in the box and took it with them. The airlines won't hold or delay the flight; if they were to, they will be penalized a hefty fine by the '*Airport Authority*'. "Time is money!" at airports around the world; at Kochi International too!
We had to hire the helicopter shuttle service to whisk them off to the airport and make special arrangement to take them directly to the *departure gate* for the flight! It was quite hectic in the backstage!

We hope (enthusiastically and very sincerely) that they are on board and on their scheduled flights without facing any problem or delay. The children would have to attend their respective colleges on the opening day itself. They were just gracious enough to take some calculated risks and entertain us in the middle of their tight schedule.
Please join me in wishing them "*bon voyage!*"

Heeding the plea, the crowd repeated the wishing: "bon voyage!", although many didn't understand what it meant.

When she turned around to leave the stage, she was peppered with a barrage of questions. A group of audience took over the very same *mic* in the back aisle which the accordionist initially used to set off this *epic saga* (the continuous tales or events). The microphone has become an iconic (emblematic representative) symbol of the *pivotal moment of drama*, and the *delightful musical instrument* at the same time!
Thus, it became known as the "*accordion-mic*!"

“What are their names?.. Where are they from?.. Where are they going?...Have you known them before or of their music?...may be seen on YouTube?...How did you manage to get them on the stage without knowing anything at all?... a bunch of lies!... ask Elise?...she should know something!.... After all, it's her show!... That’s why we came”

A pandemonium (wild confusion; uproar) seems to be breaking out. A good thing may end up as bad; the executives feared. They cannot go off the air as they have pre-announced the program to last for another hour. Ms. Bhagavathi asked for guidance, information; she questioned and consulted the executives at the Broadcast center. Well, none dared to antagonize the Television viewers and the sponsors who bought the ‘*commercial advertising spots*’ at hefty price tags. The deliberations continued for some time.

In the meantime, people began to speculate and began to offer their *best educated guess*.
“They may be living and working in Qatar; as Ms. Bhagavathi hinted”. One person opined.

“No, sir!...I beg your pardon!...it may not be necessarily true. Doha is a hub for Qatar Airways. It is also a transit center for many flights to different destinations in various countries and continents.”
A frequent traveler noted his polite skepticism.

“Then,... they may be living in Moscow. The musician was playing Russian & Romanian music and talking highly of Leo Tolstoy. Perhaps they are communists!”... “Let’s ask the chief minister!” one said laughing in a sly political parody and humor.

“Sounds plausible, though their accent was not Russian. Nobody speaks English there if the do, they will be turned in as CIA operatives, charged with treasonous high crime. It’s just not seem possible. If at all it’s true, one of them must be a diplomat. Russia doesn’t grant work or immigrant visas to people; may be to the Orthodox Serbian Slavs or others from the erstwhile Soviet Union. You may find some Kazakhs,

because the Kosmodrome and rocket launching pads were there in the Soviet era; also, a few minorities like Tajiks, Uzbeks, Tatars, and Chechen who are brutally discriminated and awfully treated like the Blacks, Hispanics, Asian and religious minorities in the US".
A gentleman summarized his extensive knowledge of his world view.

"Well, listen, Leo was an ardent anti-communist!" Another tried to discredit the premise and settle dispute once and for all.

A college student interjected: "yeh, how about Austria, or Germany? They were playing Beethoven and Danube. Danube flows from the Alps, through Southern Germany, and Austria! May be through Ukraine too. My cousin Manilal who studies in Odessa often talks about the Black Sea where the river ends. Yeh, what do you think folks?"

" Yes,...yes...they are all speculations without evidence. Let's wait to hear the official version!"

Ms. Bhagavathi returned after animated discussions visibly agitated. She said:
"Everything was arranged and orchestrated by the Broadcast center, without any knowledge or participation from any of us in this hall. It was quite a surprise '*extemporaneous*'...ad-lib... unscheduled; just added as an '*adjunct*' (something added) to the program. We, here at this location simply followed the instructions from our bosses. That's the *protocol*! (A code of correct conduct); "yes, sir" is always the standard replay by the subordinates; implicit obedience is the basic corporate curtsy, you know! It's inherent, and implicit! That's the way it is, Ladies and gentlemen!". She said in unrestrained exasperation.
Immediately, Ms. Bhagavathi realized she was too cynical and critical; perhaps, displeasing her superiors who are always too egoistic about their '*decision-making genius*'.
She hesitated a bit that she may have displeased her supervisors; then added: Forgive me, please!
It is just the standard operating procedure practiced everywhere!

I was told that they, themselves, received the tip and recommendation from an anonymous caller. This person heard their music, and the particular song by Elise so beautifully played at their home often in the evenings.

His sincere and original intention was if it were somehow possible, his tip might help to give the concert-attendees a rare opportunity to hear them, especially on this auspicious occasion of the grand opening of the Art Center. "Wouldn't it be fantastic?" he kept on asking the studio managers in pleading refrains. He imagined they have to play just one song and done;... only the Elise's song.
The studio was also made aware of the time constraints, tight schedule, urgency, and how critical the times were.

I do apologize for all the mistakes and misunderstandings!"
"The bottom line now is....Did you enjoy the music of the evening?"

"Very much!...A lot... excellent!" was the unequivocal responses.

"Did you like the unfathomable mystery?"

"Yes,...Not so much..., Somewhat...., everything OK, now, Miss.!"

People expressed varied responses with some ambivalence mixed with conciliatory remarks.

"Did you have a good time?"

"A wonderful one,... yes indeed...Elise;... the sound, music;... the beautiful hall...yes we enjoyed it all!... quite stunning...in spite of the puzzle and mystery!"

"Would you like to come back and listen to us
if we arrange another guest appearance with the family musicians?"

"Gladly! ...most definitely!...They are a wonderful family!...God bless them!...Kind regards!..."Please find out more about them...and promptly post it! We look forward to hearing about them!"...Ok, Ms. Bhavani?

Elisa took to the Mic; said in a much relaxed tone:
"We are sincerely thankful to each one of you for being such an appreciative, attentive, and interactive audience. You bestowed great praise and recognition for our performance. You're all very special. We are very, very grateful, indeed!... Enormously! You are all very good sport! You've expressed yourselves freely in good humor, support, and kindness; We'll keep you close to our hearts; hold the music and the songs you heard likewise! Live a good life, do good deeds; God bless you, all!"
Have a safe trip home!" she shouted in an after-thought.

It all ended! The enchanting evening concluded in intrigue, style; surprises, and suspense reminiscent of Sir Alfred Joseph Hitchcock movies. But, this one though, without any crime committed!
None got hurt! No weapon used; no blood spilled!
Such a glorious ending! A nice plot for a movie!
Indeed! Some said in affirmation.

People found an affinity and a '*sense of belonging*' to the place and music as they have never felt before. Laughingly, and jokingly they formed a line and let other people go ahead and pass in front of them in unusually seen courtesy and politeness. This is quite unlike the melee at the bus stands. People have instantly become refined and accommodative like never before. In a reverential, magnificent environment and settings, one has to behave in such order and decorum!

There was no rush or stampede: no hurry, scurry, or scrimmage. All quiet, peaceful and civilized conduct; demeanor and behavior. They smiled and hummed tunes; found an uplifting spirit; a jovial jolly good mood; joyfully dispersed and head home in fine cheer and pleasant temperament.

The melodramatic events of the evening has awakened curious interest, senses & sensibilities, as well as certain faculties of the mind by the *sensorium* (apart of the brain regarded as the center of all sensations) of the resourceful human brain. Everything in the universe requires a trigger or stimulus to launch itself into action from its *quiescent* (inactive) or dormant state. You may call it the '*universal law of defaults*'. A noted philosopher in the crowd thoughtfully commented.

In spite of the buoyant atmosphere, they felt that both the fleeting time and the guest musicians left without saying "*goodbye*". Somehow, it hurts. Most are curious to know whether it was a clever show-biz '*setup*' or not. If it has happened naturally, or an incredible coincidence. However, this *riddle* must be settled for certain; it must not linger on obsessively as a total mystery. We must find out the true story!

The audience liked to know them better. Their contributions to night was very impactful and showed new dimensions to showbiz entertainment. Their presence and performance have been quite inspiring and significant. They were not just any family who liked music and played it well. They possessed a certain charisma and pleasing mannerisms in their interactions, and expressions. There are certain things to be learned from this.

The experiences of the evening extended human senses further and wider. However, the confusion is still quite perplexing. The suspense rcmains suspended in the air inside and outside the hall. A relentless obsession with the mystery that is left unsolved, looms; preoccupies the minds: continues to nag minds.

A certain possibility that it may be resolved later in the normal course of time and circumstances, offers little consolation right now! Certainly, it should not tarnish any of the marvelous things that transpired here this wonderful evening!

All is good that ends well !

Epilogue *(afterword):*

"We made a great discovery that humans are born with sagacious minds. It needs stimulation to dive deep and probe its depth. It is very fertile! What is sown on it, grows and multiplies; gives returns in many folds. That's again the universal law! Ask the farmers! But, one must carefully understand that it gives out only what is sown, not anything different. That is to say if you sow good things, it returns the same in abundance. If you saw rice in your field, it returns rice; on the other hand if you saw opium, or marijuana it gives the same thing in higher quantities. Unfortunately, the same fact is true for stray seeds and weeds of temptations, habits, bad, and malicious thoughts; disposition for violence, revenge, anger, pessimisms, depression, etc. Inevitably, both positive negative thoughts enter the minds needing none of our conscious permission, via intrinsic (belonging to a thing by its very nature) propensities (natural inclination or tendency) of the mind, or extrinsic (originating from the outside, external origin) influences, by false & malicious propaganda, deranged imagination, stimulation by crime themed cinema, Television, music, media, social interactions, conversations, events, news-stories etc. They may be regarded as purely '*realistic*' as what is happening in our daily lives and society. What is bad is to sensationalize and dramatize it for profit, revenue, fame, and influence. Such stimulations predominate and proliferate in recent times because of its sensational aspect, and monstrous marketing power. The '*evils*' will have to be weeded out, before it gets a chance to take root, grow, smother, and destroy us. It's a continuous process of weeding, detoxification, and sanitization!

We need to know a lot of things in life. In order to discover them and to apply them judiciously on a daily basis, we need to have education. Education gives us the capacity to learn, a level of understanding, and enlightenment. It takes diligence, sternness of purpose, discipline, focus, self-control, and determination to achieve the goal. A clever way is to engage our minds in creative ways; setting goals, achievement levels, and excellence. We have to persist and persevere to cultivate useful skills, arts, hobbies etc; practice contemplation, meditation, introspection; exercise body & mind;

follow a balanced mental, intellectual, physical, and healthy routine & lifestyle. This new Art Center is one such center that promote '*Humanities*' (The study & learning of language, literature, arts, philosophy, etc. that are concerned with human thought and culture). *It's a monumental gift and rich resource for human advancement!*
In addition, we must devote enough time to ourselves!

Let's skip some '*overs*' in cricket matches; let a few balls roll over the '*boundary line*', and let few wickets fall unseen, and '*live*'! Let's not chase all the stars, stats and '*wicket stands*'; leave some silly political drama and circus to themselves. Let them argue, vilify each other, and make fools of themselves in public. Let them profit or perish from it; nothing is in it for others to learn, or benefit from; all, just *humbug* (nonsense & rubbish) and mere waste of time!

Let's do something creative and perform, instead of watching other people performing. The stars & performers are amply compensated for their narrow specialized skills. Only time will tell what other important things in life they have been missing all along. Perhaps, they omitted learning; neglected family time in bringing up children, their education, participation in family activities, so on, and so forth. It' ll be all too late, later when the reality dawn on them and prevail, inevitably! It takes much more effort and abilities to be successful and productive throughout life, than mastering a single specialized attainment and succeed for a tiny spell of time. Let it be known that life is not a single skill pursuit!
We don't want to be specialists '*who know everything about something and nothing about anything else*'!

Let our motto and belief shall be "*man's reach and capabilities are bigger than his grasp!*" Let's do what is important that's enduring and valuable to ourselves; benefiting the whole humanity!"

This was the theme & subject matter of discussion and musings of a scholarly (having the qualities of a scholar; concerned with academics) family, at the dinner table shortly after returning home from the concert.

It turns out to be a motivational and, inspirational session; served as an immediate and rewarding collateral benefit from the concert experience!
In the meantime, in the background, the Television was broadcasting live a D/N ODI match between Australia Vs. India from the Eden Gardens, Kolkata.

The mother got up and switched off the Television.
The children, retired to the study room and started doing their 'homework!'

Today is the 60th Birthday (Shastya-Poorthy) of the head of the family, Shri. Tharamel Hariharan Thampi. The concert was a *treat* for him on this auspicious day in life, arranged by his eldest son working at Google abroad. Traditionally, it is the time, a person retires from his profession or career he has been tirelessly pursuing to make a living. The next day, he will be entering into "*Vanaprastha*", the penultimate phase of an active life and gradually detaching himself from the daily grind, material wishes, and pleasures. He is availing this opportunity to pass on his legacy and gathered wisdom of the yester years to his beloved wife and children. He is entitled to an attentive hearing that his family has now respectfully obliged him with!

It will be quite interesting to find out how long the newly found spirit and sentiment will endure, and if the children really improved up on their grades.

Only time shall tell!

............A Few days later

There appeared a posting in a typed letter on the network's website.
It reads under the heading: Follow-up comments on "*The first Concert*"

"At a certain part of the concert, our network received a tip and a strong persuasive recommendation from an anonymous but an earnest caller. This person heard the music, and the particular song "Just say, I heard you!" by Elise so beautifully played at their home in the evenings. He kept on saying in great excitement and exclamation "one should definitely hear it. It will be a stunning surprise to the Television viewers and everyone else!" Now we know that he was referring to the "Family trio of musicians". The caller also said they were doing a video shoot at their ancestral home with "The *Great Chamundi* Chenda Vadhyam" company and the instruments the guests brought with them for the '*shoot*'. The whole villagers were invited. During the shooting, we sat on the fences, on the ground, on the terrace, verandhas, and leaned on the arch columns of the all-around-porch. We were served with sandwiches and lime juice. It was great fun. All of us liked it very much, as we were in the video. The accordionist was also directing the show. There were 4 camera people with big TV cameras. We used our mobile phones. The school children also did participate. They sang a song 'Soorya-Namaskar' taught by the guests last year when they came to sing at their school alternating with '*Vande Matharam*'. They filmed some 'yoga' routines as well.

He thought it would be fantastic idea for the occasion if an appearance by the family could be somehow arranged and give the concert audience a rare golden opportunity to hear them play the song together with Elise. There is no chance better than this amazing occasion of the grand opening. Please go for it, please!

Hearing the unusual music in the evenings, he and a bunch of his friends used to go to their courtyard to listen to. They were cordially greeted with hot coffee, and refreshments. Each time they added one or more songs. These folks were extremely impressed with the hospitality they received and the beautiful music they heard.

One of them, who owned a private cargo transport service, was entrusted with transporting all the luggage to the Kochi International Airport. The story goes that it is he, who first informed the family of Elise's concert and the grand opening of the spectacular Amphitheater. Persuasively, he told the visiting family that it would be a great opportunity and great experience for their children to see and witness the grand ceremony. When the family said that they were 'time constrained' and didn't want to bother with standing in long line for tickets, his crew promptly offered their own tickets to

entice them and make them change their minds. The family ultimately agreed. They were given ₹ 4,000 in return, to buy tickets for themselves.
The conversation originated immediately when the gentleman stood at the pedestal microphone and sang back, the chief Studio executive took the call. At first, he refused to entertain the matter any further. It was summarily rejected on the excuse that the orchestra has no accordion to play on. The caller persisted. He consulted the brother who was riding with him. Then, promptly stated that an accordion is among the luggages, and he could bring it through the backstage. In the great excitement and amusement created by the occasion, a decision was suddenly made at the highest level. We must admit that, at the studio, it was an all consuming celebratory party mood and frenzy. Everything was going extremely well too. Everyone wanted to be a good sport and thought it was a terrific idea; unanimously approved the proposal and gave the 'go ahead'. We didn't have the slightest idea how this risky gambit is going to pan out, but did heavily rely on the ingenuity and resourcefulness of the well-trained Production staff at the Art Center. Ultimately, we simply seized on the presented opportunity! Nobody did express any concerns, only excited about the potential surprise effect it can make and boost our image.

Apparently, the family nor the Art center staff knew anything about the plot, nor did any other as nobody did know in advance how it may turn out. We didn't want to share any information with others, as an immediate decision was essential to pull this out smoothly and effectively. Our sincere apologies! That's exactly what has happened and how it happened.

At present, we only know that the family came for an occasional visit of the ancestral land of their parents and grandparents to give their children some understanding of the culture and lifestyle of the native country. They stayed in a 'home stay' resort for the final week. The Resort manager refused to divulge any details, respecting and citing the privacy and confidentiality policies of the establishment.

This will conclude our 'search & find' efforts in this matter. Respectfully, we ask for your continued support and understanding!"

Sincerely,

Maha Deepam Kairali Television & Entertainment Network Ltd.

Apologetic Elise

" I'm remiss in not asking names, in spite of the fact that I had curious personal questions myself. I had been searching for confirmation for my own suddenly occurred beliefs which were growing strongly within me by the moment. They were even tormenting!

The concert night was full of serendipities (The occurrence of making fortunate discoveries by accident). Please understand that I was literally caught in a web of intrigue & surprises, "contiguous events" (connected in space or time) of unscripted and bizarre occurrences. A lot of commotions and confusion prevailed everywhere. Everything was 'the first time' and that contributed further to add to the usual anxiety in a tense situation like this. I was confounded and confused. It diverted my thoughts and attention.

I forgot all the common curtsies at the moment under the grip of emotional disruptions.
I was also trying to reconstruct in my mind what happened to me many years ago. I was showered with flattery, not by fleeting words, but by significant and meaningful actions; the pieces of music they have chosen were so touching and surprisingly relating to the most memorable events in my life! It was extraordinary,... befitting...timely...stupendous...everything that has happened was totally incredible!

I witnessed a taste of the same flair and flamboyance, wit and wisdom in the face I was intently staring at.
I was transfixed and taken to a scary world I faced in my childhood.
Frankly, I still have to recover from it and regain my present self!

Later, I regretted, when I had a real chance to review the video recording at the studio, that I was rudely disrespectful of them. Each one of the family was noble; conducted themselves admirably; showed refined mannerism. We enjoyed their grace and talents; yet, treated them in ungrateful and undeserving ways. I do apologize!

This is the 2nd time, that I failed. I'm mortified!
I apologize to them, and all of my fantastic fans.
I know, they are so dignified, broad-minded, blessed with big, kind hearts.
They will be surely magnanimous to forgive me.
I am hoping they will be in touch with me some day, sooner than later!
I will be also making efforts to make contact with them. If they are anywhere in the Gulf, I shall visit them in person during my upcoming concert tour with the band.
For now, I send them my warm regards and the very best wishes"

- *Elise*

From the scenic 'R-Block' Kayal Resort…

A video documentary appeared only a few days later on the same Website, in response to numerous inquiries and comments from viewers.

"The family spent their final week, at the R-Block Kayal Resort, on a large section of the Vembanad lake, a well known place for tourists and houseboats where they stop by for refreshments, food, bird watching, and 'walk around'. They chose this place because it was close to their hearts. Their own ancestors were pioneers in the efforts of its reclamation, converting it into vast fertile agricultural fields feeding a few million people; caused a resurgence of the region; became vital to the economy and food self-sufficiency of the state.

It was also the biggest and the most adventurous human endeavor of the time!

The visiting family of four spent the days enjoying canoing, fishing, and swimming across the canal separating sprawling blocks of rice fields. They liked to drink tender coconut water; eat fish caught from the freshwater lake, by themselves. They were very curious and active; enjoyed the natural surroundings.

A gang of village urchins used to follow the kids around and mimic their actions. They seemed to know the most about them. Eloquently and excitedly, they acted and babbled out to us many hilarious stories about them.

In their own simple, plain words:
"We watched every move and action made by them. Sometimes, we go and play with them, run around the place, help them catch fish. We know all the good spots in R-Block. The young lady was often seen eating the "Kathali" berries making her mouth dark pink. We saw her boiling them and squeezing essence out of it. She smeared it on her nails, lips; put it on forehead as the "pottu" (beauty 'dot-spot') and on earlobes".

“My mother, seeing it, sent her some saffron '*Kumkumum*'. She put it on; put some on her own mother too. When they left, my mother gave them a small jar full of Kumkum. They paid her ₹ 5,00, although it was meant to be a gift”.
Vasu said, holding his bamboo fishing rod. A live bait dangling from it wiggles hard to extricate itself.

“They were very excited at seeing the birds and the passenger line boats plying between Kottayam & Alappuzha, also many tourist boats & houseboats pass by in between. The boy and the girl would run along in pace with the boats and warmly wave to the passengers in them.
We ran behind, joining the fun. They were happy kids!”

“ Dasappan, the toddy tapper offered them fresh Toddy right out of the trees. But, they pointed to us. They said they shall never drink alcohol”
“This one, Karanan, swallowed the full bottle in one gulp without sharing a single drop with us. He fell down on his way home...hah, hah....ha ha haa...yes...this '*Kutti Kudiyan*' (the little drunkard)”. The village boys said in great amusement and taunts.
“As they were leaving, they gave away some '*goodies*' to any passerby; toys to children. I got a face mask looking like a “*Bhootham*’ (a dreadful monster) and a plastic flute;
The boy showed us how to blow sounds. He played a music with it before giving it to us.

“This guy got a hat;... hey you! Where is it now?!”
The big bully asks

“He thought it’s a small fishing net used for catching small fish for baits; doesn't know what to do with it. I showed him what it is meant for!”
Starting fist fights with one another. The cameraman skillfully evades a left jab that flies past across him.

“They didn't speak any Malayalam; only something else; talked to us in “*Kadhakali*” style- just only in signs with hands, mouth, and eyes. It was very funny. We laughed a lot!. .... oh.. oh..haa...ha...!”
Each, one by one, showed some “*Kadhakali mudras*” (gestures and facial expressions as performed in the 'pantomime') recently learned in some group lessons at the temple.

“One day, we followed them to the far corner. They sat on the jetty with their funny looking fishing rods. It had a wheel on its rod. They used a long nylon line with a big '*anchor hook*’ on it; had only some plastic baits, no live baits. The fish liked the different colors on it; they thought they are some new kind of small fish and will taste good; took a lot of bites. The kids started shouting loud in great excitement. We told them to keep quiet and not scare the fish away. They said fish cannot hear their shouts; only vibrations. They caught many -Koori (catfish) & Kanambu (freshwater Mullet); some Karimeen (green chromide) too. We borrowed a basket from the tea shop nearby; carried it for them. Their mother gave us ₹ 10/- each. We went back to the shop; bought and ate a lot of Appam and curry, Vadai, and tea with it, until we were too full and couldn't get up and walk. The nosy server asked us where we got the money from. We kept quiet all the way; pretended fainting on the desk. After sometime, he shut up by himself!"

"In the morning, they jumped off the boat jetty in to the water, stayed underwater for a long time. We thought they drowned and shouted for help. Later, they came up far away and swam across to the other side; sometimes they swim like us (Freestyle); sometimes on their back (Backstroke); sometimes do like a frog, crawl, and kick (Breaststroke); I liked it. They could glide like a fish (the boys demonstrating the glide holding hands in 'Namasthe' sign); also up and down like a 'jumping calf' (Butterfly stroke).

"Sometimes they race; the father always lose; reaches last; the kids stretch out their hands towards him; he pays ₹ 10/- each for losing!"

"But he gets it back though, in the boat race; the children cannot steer the boat straight!"
A sympathetic lad adds.

"The mother brings banana to the jetty after they finish swimming. They take only one each, and distribute the rest...they were "Poovan" variety from the front yard; big, and sweet, grown right there (pointing to the spot)!"
All start pointing to the same spot; hands colliding while doing so; reason enough for another fight.

"The girl was shy; the boy taught us diving and said if I do well, I can go to Olympics; I'm now practicing for that;... every day as they did;..maybe I can meet them again...at the Olympics!"

"He said, then I can win Gold, silver, or bronze; I want to win Gold only; give it to my "*Muthachi*" (great grandmother); she always wanted a gold necklace, but never did have one. When she asked for one, '*Muthachhan*' (great granddad) scolded her;told her off; she says. I feel sorry for her. Now, I can do forward and backward flips as good as the boy"
He continued excitedly, as if he had already won a gold!

“He is not that good; I can beat him; he gets tired fast!”
His buddy Sashi commented with some contempt and jealousy.

'Nee, poda! (Go away, you silly)... Shut up!....*Mandachare*!, (you damn fool!)...
I am talking about diving; you always land on your pot-belly; turn it black and blue; you learned nothing; you'll always be here; drown one day.
I'm good; I will go, you..get lost!”
It seems to have irritated the aspiring Olympian for disrespecting his beloved great-grandmother. The urchins quarreled back and forth; leading to some physical altercations.

“ He thinks, his Muthachi will gift it in turn to his “*Mura-Pennu*” (designated future-wife). You know, ...Sumitra, his aunty's daughter, his cousin, is his 'betrothed'. That's their custom. He is not permitted to talk to her until after he marries her. He is too shy and timid to give it to her directly”.
Some are discovering the top-secret only now, with this revelation. They exclaimed: “*Am-ba-ta, Kalla”*! (you, devious coward!)”
Continuous laughter and taunts go around for a while bullying the boy, followed by rounds of loud laughter for each repartee (quick and witty exchanges).

“ Sometimes, they run along the bank, from one end to the other; the children had to stop every often for their father to catch up. One day, he slipped and fell into the ditch..ha ..ha..haaaa...covered in mud all over his face; oh! That poor man! Then he jumped in to the canal to wash it off before going home totally wet and draining water all the way”

“ One day the kids decided to run around the entire R-Block. They said they were running a Marathon. Some of us ran behind, but when tired, returned to the spot and waited. They were not to be seen for a long, long time. We told the parents; they got alarmed; the mother panicked, started to cry; sent people to go and look, search and bring them home. One group went this way, and the other the that way. Noon came, when we saw them slowly limping along, brother holding his sister's hand. Both looked tired; sweat streaming down; could not speak a word”
The boys narrated, acting out many scenes from their funny recollections.

“..No! They smiled at us; don't you remember?”
Corrected one among the unruly gang.

“ The boy always guarded and protected his sister; often told her what to do; she obeyed most of the time; ...they never quarreled like us!”

“I never quarrel with my sister; she always listens to me!”. One boy claimed, which started another round of squabble (a petty quarrel) among them.

“One day, they gave us cake, very tasty…' better than ‘*vatteppam*’ (a fluffy local delicacy, sweet and delicious flour recipe with yeast and a pint of toddy). We have never eaten one ever before or after.....this bulky one, the greedy Gopi grabbed two pieces at once with both hands,....you fat pig,... go away!” One of them, say in disgust, starting another push, and shove.

“Ah, you picked the biggest slice...how about that, Maniya, you selfish?”
They continued arguing back and forth for sometime about the same subject in funny facial contortion for each remark pretending to pinch the other’s earlobe.

“They collected fallen coconuts; heaped them up; climbed to the very top of it, sat and rolled down to the ground landing on their bottoms with the collapsing pile, 'laughing all the way'!”

“When the tree climber crawled up the tree, the girl was scared; closed her eyes and prayed until the man came down to the ground safely. The boy picked up all the scattered bunch of coconuts and heaped them as before and had fun with it; sometimes we help them to pile them back up;...he gave us one coconut each. He said the property belonged to his younger uncle!”
The cameraman asked: “does their uncle come here frequently?”

“Yes,..yes to harvest the coconuts” together they answered at once.

“Sometimes, with houseboat also” another added.

“Can you ask him where the children are now?” the photographer asked

Silence!…The urchins looked at each other curiously.

“He doesn’t talk to us; ...we don’t ask him also” one boy said cautiously.

…“When they left, we all cried; They were fun; they looked very sad too...waved to us... we shouted ..Ta, Ta..!; “May be they will come back again...next year!” Some wished.

" When we went out fishing at night, we heard their music. Every night, we caught nets full of lobsters & crabs. Our kids and everything else are quite, and gloomy around here now".

"Damn!.. Even the fish has gone away!"
Some villagers were squatting on the grass mount under the shade of the banana plants playing cards. They were gulping fresh and tasty Toddy out of bottles with long narrow neck. Said in slurred slang of words and in hopeless tone.

"The resort is now closed. Nobody came here since!
No one seems to know anything more!
Some other people we asked, also said they missed them very much.

That's all folks!" The reporter said in his summary.

- Radha Krishna Panicker reporting on-site, from the magnificent R-Block Kayal Resort

A bottom crawl runs by announcing the sponsorship of the video report:

This report is sponsored by your favorite the **CRUNCHY, TASTY**

Barley Biscuits Ltd.

End of program

....

The local villagers offered the Television crew Toddy in bottles, and dishes of 'Tapioca' in flavored coconut gravy showing customary hospitality and welcome. Drinking is prohibited on the job and on camera; therefore politely, but reluctantly declined.

No sooner than the manager left the scene, the rest of the crew started shutting off all cameras and collapsing the tripods; packed them up in hard metal cases. Once everything is secured, and put away, they sat down with the locals under the shade and enjoyed several sips and bottles of Toddy, merrily cracking jokes and sharing stories with the hosts.

One member of the Television crew checked the latest postings and read out for the benefit of others. It read:

".....if no one can find and locate them, they are mythical characters!...ha.... ha. Haah..; add this too, to our already existing long list!"

" I'm puzzled!. Who is the real hero?; The Arabian Sheikh or the designers & artisans?; Elise or the family singers?; Tolstoy or Beethoven?; The '*Manthravadi*' or '*Ageles*' Mathappan; the author or the accordionist?; the siblings or the urchins of R- Block? It's very interesting!"

"Anything at all about the "family trio?" a member of the production crew asked.

"No, nothing at all... Well, that was the latest comment, folks!"

Addendum:

The art center has been getting rave reviews, so also the First Concert at the grand opening gala. It's regularly on the news media and conversations for one reason or the other. It has become a major focal point of widespread interest. People continue to report what they know and experienced in the participation of different events, tours, and visits. Several readers in the professional fields explained many aspects of information given in the book. They began to appear on the website.
Some interesting ones chosen at random are listed below:

"The description of the musical ending by Elise is so brilliant, tremendous writing!"
Michael Vincent, Dadar, Mumbai

"I was actually fascinated by the transformation of Mathappan, the comedian from 'Mathravadi' Sorcerer to Angeles Mathappan. It's so hilarious!" **Vinny Atkinson, Oshkosh, WI, USA**

"You know! It has many different subjects discussed as an expert in each one. I guess it's written by at least three people. If you look at the author's name, you see three names joined together- Aurora, Bobby & Alice. All girl's names. That's what my wife says. What do you think?" Joe Jacob Mathews, Evanston, Illinois.

"My wife says it touches her deeply, I hear her sighs when she reads the poem or hear the song. She is Elise's godmother and aunty. The incident did really happen. Nobody saw. She, nearly drowned. Celine, her sister, had told her so!" *Shiju Jacob, Moovattupuzha*

"A piece of great musical appreciation, written and explained in many details. I recommend musicians to read and get some insights. If you do something, you should do it as best as you can; that's my thing! By the way, we've some decent music books. Please check it out!" *Bernard's CD & music books, Panaji*

"...versatile & in depth, although I didn't understand quite many of it; I enjoy music better now, the same with my 'Bhagavathar' (music teacher). We will go and visit the center soon. I heard they give tour orientation on Thursdays. When I watch orchestra concert on TV, I watch the conductor's actions also" *Seetha Kumari, Chenganapally*

"Starting each chapter, I got lost thinking that it is not at all relevant to the theme, but towards the end, it blends in with the story in brilliant pertinence and significance. Unique and splendid writing style, I must say! ... Its educational, motivational as well; most part, reads like a bewitching novel. I don't hesitate at all to award a rare 5*****!" *Gopala Krishnan Nair, Kulangara, Literary critic*

" The Prof. of Telecommunications at the Engineering college, Kulathoor verified:
'They are all real, applied technology, technically sound in all respects, very educational'...recommends students to read it to get some practical knowledge and good tips" *M.Krishnadas, Nalanchira*

"....Right! it contains valuable information for musicians, and many others. You will come to know many new things. I like it very much. I also recommend the book!" *PothanMaster, Mannar*

“.....the contextual explanations were quite useful, though many times, complex and difficult to understand in spite of the fact that my major subject is English Literature & Linguistics!”
Ms. Daisy Antony, St. Theresa's College, Ernakulam

“ It is an excellent book to expand your horizon and knowledge; also wide-ranging. A macaronic composition as well with a number of Malayalam, Latin, German words & phrases, in addition to many vernaculars of musical, scientific, engineering and computer terms. It has a lot of sense & substance to it. Enjoyed it very much, many kudos!” *Prof. Thai-cherry, St. Berchman's college*

“a tedious, encompassing content, but, anyone can find many new facts about music, also relating to other topics of individual interest, lifestyle, and careers…in those respects quite remarkable. I appreciate many of the descriptive facts on music & the musical instruments; Very interesting *****!”
Enoch Erasmus, music director, Richmond

‘It was exciting when the brass ensamble checked in with the family players. They played well. I don’t know if they knew the song already. The concert was so very fantastic” *Gigi Ann Lonappan, Trichur*

“ Is translations available in Hindi?. English is a bit difficult for me!” *Kamala Malhotra, New Delhi*

“It is so touching to my heart that Zabrina has become the father's most loved daughter. I have only daughters. Four beautiful daughters. Remember her birth was the only reason to say triple **talaq**, although the husband reversed it later. I thank the author for writing such a fantastic story.” **Fareeda Rizwan, Lucknow**

“Yes, in the conversation between Sheikh and son, it says the following about the law of gender determination.“It is inviolable; cannot be altered by any human force, campaign, court authority, legislation, decree, or thalaq!” I think we should teach the youth these things. Then we will have fewer problems in marriage and family. Amina Nasserudheen, Kuweit

“by the way, triple thalaq is illegal in India after the recent supreme court ruling. But, the husbands always re-marry and carry on. Poor women are confused about their status if they are married or not. Again, th poor women suffer, as always”
Noora Mohammed Beevi, Nagpur

“It is formless; uses a number of metaphors. The author drifts a lot. You have to wait to the end to grasp the substance and the plot, also the convoluted series of mysteries. Quite fascinating though!”
Sivarama Swamy, Music critic

‘I like the redemption of Angeles Mathappan. He is my neighbor and college mate; has been very witty from childhood; very funny; imaginative and gifted comedian...ha ha haaa!. By the way Cherai beach and its description are very nice” *Thomaskutty, Kurunnelpady, Koratti*

“I gave it to my children in Bangalore. They are into music, violin & Guitar. They should find it interesting. Author is right, we must include piano and accordion in our music. They are nice to hear”
Omana Pillai, Haripad

“Personally, I liked the suspense at the end, remembered the same part in the 'Sound of Music'. During the award ceremony, the von Trapp family fled to the surrounding mountains on their way to Switzerland, a neutral country, escaping the secretly waiting Nazi soldiers who wanted to forcibly take Captain von

Trapp. Previously he had boldly declined and defied Hitler's commission in the Navy. Every year, during Winter holidays it is shown on TV. Good job! Nicely written and interesting. Best regards!"
Ms. Bhanumathi Bhaskar, Orlando, Florida

" Yes, the Jews suffered a lot!; appreciate you mentioning it. It was quite moving to read that part. Also, Long live your King of Kochi! I shall help you reconnect with your 'long lost' friend Basil. I shall look up in our registry and let you know. I'll also mention to him about you. Please give me your contact details and the year Basil left Kochi!" *Rabbi. Efrem Jousef Lieberwitz, Tel Aviv*

" I learned more about the elephants whom I see every day, and yet not known. The mythology of God's own country is very, very nice. I felt I was on the beach that day. Many, many thanks! Sir/Ma'am!"
Prof. Ambujakshan Ambadil, Konni

" To be frank with you, I choked a number of times, Avarachan too. Our teenage daughters loved it; believed everything is true. They will ask their Facebook friends & contacts about the family!"
Alamma Avarachan, Aranmula

" The main purpose is well served. It was a sensational concert, one of a kind. My family thoroughly enjoyed it. We sang together and loud, first time as a family; maybe we got inspired by the family musicians. We want to do it more often too; look us up on Youtube soon coming!"
Damodaran Potti, Trippunithara

"We need to encourage our music masters and directors to try out new things; add more instruments, varieties, and styles. I like the sound of accordion much better than harmonium. A.R Rahman has been quiet innovative; so also RD Burman to some extent. Now, it looks like the same spirit has disappeared. I'm not sure if Piano is suitable for our kind of music. We can certainly adapt it. I think we should try it, though you need a Lorry to transport the large piece. Then again if an elephant can be transported, any other thing can. Hush! Just trying out a joke for my sermon, pardon me, please!"
Fr. Mamala Muthu S.J. Chennai

"The singing in the bathroom to prove a point in acoustics is absolutely a wonderful imagination by the writer. I tried it out. It is true. It took a longtime for me to finish the bath as I enjoyed my singing tremendously and did not know when to stop until my wife came shouting." Zacariah Philips, AR, USA

"You know, most novels have one plot. This has an interesting one in every chapter. All contribute to the main story. Just super interesting when it comes to the final pages! Ian Bancroft, Glasgow.

"The most powerful paragraph for me is the 'family musings' at the dinner table. Every person, especially the young should read it. It is a teaching moment, well stated with profound meaning. All along, I have been trying to tell the same things to my children. But couldn't find the right words or temperament for it. Now, I can simply tell them to read it. I like the mention of my native Sri Lanka!"
Pastor. Bartholomew Balthazar, Galveston, Tx

"The auditorium is so beautiful to see; the music sounded heavenly. I didn't know how to use many things found in my seat. My wife Amina is an electronics engineer. She explained many things in the book. We both liked the parade very much and its explanations. Very nice! I am looking for the next major event to attend. It is really a great experience! I might buy a home theater system soon, but the damn thing is very expensive!" *Abdullah Latif, Kozhikode*

"There's a Qatar Philharmonic hall situated at Katara Cultural Village in the West Bay. My husband worked there as an Electrical Engineer. There are a lot of Expatriate Westerners living there. I guess the family trio may have some connections to it" Colleen Guilford, Birmingham West Midlands

"The cross-section of readers are many; something interesting for everyone is surely there, as information, teachings, history, mythology, science, documentary, novel, modern & futuristic, rustic & urban, arts & entertainment, or statements of moral lessons. It can engage many minds, not only musicians. It's wide-ranging and extensive, technology heavy and thoughtful. An excellent work. The family singers played their instrument so well and with confidence, also showing their true emotions and feelings!" *Piush Publishing, Mumbai*

"Manthravadi to Angeles by the decree from cardinal, His "Beatitude and Eminence", is so hilarious, mei man! Please include a cardinal's cassock or a vestment with his red biretta" *Brian Maynard, Notre Dame IN.*

"the singing in the bathroom and while bathing in the river are really funny pieces of descriptions, but wonderful and excellent examples. I thoroughly loved it. Adi poly Thanne, Mashe!" *Joy Tharappel, Ernakulam*

"Thie music commentary is tremendously enjoyable. It shows many more ways music can be understood and enjoyed. I read a lot of music books, none has described it so beautifully. It is entirely interesting, just fantastic. Has anybody heard music by this writer? " Rohan Reddy, Hyderabad

Citations & attribution:

https://www.genome.gov

http://EzineArticles.com

https://www.nasa.gov/

https://commons.wikimedia.org/wiki/Category

http://www.thefreedictionary.com/

https://images.search.yahoo.com/search/images

https://www.pinterest.com/pinterest/

http://www.pbs.org/wgbh/nova/

NOAA

National geographic

Television / Wildlife/ BBC /free documentaries

Text books

http://online.umd.edu/courses/

Encyclopedia Britannica

Wikipedia

wikianswers

Cover design by: Aurora Bobbyalis

Book Information

Published! **Available** globally for sale on IngramSpark Content Group's 40,000 retail and library partners, both in person and online!

The First Concert

It's about a magnificent concert taking place in an acoustically & aesthetically appointed, and splendidly equipped grand amphitheater on the occasion of its grand gala opening. In its course, several stunning incidents sporadically pop up; become an integral part of the proceedings culminating in intriguing drama, gripping mystery, and nagging obsession.

Notes :

1. *The author reserve the rights to edit, add, alter, modify content as appropriate without notice!*
2. *The online readers are advised to ensure that their devices can access and download the specific file formats!*

AVAILABLE IN 3 DIFFERENT FORMATS

1. **Print** Book Order Info — **ISBN: 978-1-0880-4387-5** [Trim size:..." *..." **Print**]
   Pages 280; *Picture collages (40)*
2. **E-Book** — **ISBN: 978-1-0880-4395-0** [Format:-**EPUB**]
   In all popular ***Ebook*** formats
3. **Download**Web site — ***https://www.desi-paradesi-diaspora.net/***
   Digital download in compliance with DMCA"guidelines
   File size (Download) **21.85** MB; 24.00 MB PDF;
   Word count: 81,899

Publisher Bodoll mm LLC
Book: Hard copy, E-book; global distribution
Kindle, Nook, IPAD, & Digital Publishing & "*Download from website*"
Author AURORA BOBBYALIS
Key search words :Book "The First Concert by Aurora Bobbyalis"
ISBN, Publish date August 14, 2022
Distribution & Availability: **Global** distribution by IngramSpark retail network
Digital download in compliance with "DMCA" guidelines

Preface/Synopsis online, free reading
Chapters described under "Synopsis"
Glossary English
Native words
Index provided
Barcode
Distribution & Availability: **Global** distribution by IngramSpark retail network

*Other **Books** Published::*

The Biggest Regatta, Kerala Vallam-Kali

Author AURORA BOBBYALIS
Book Pages 400
Picture collages (86)
File size 155 MB; 420 MB 600dpi PDF;
Word count 133,525
Publish Date 01, Dec. 2021
Publisher **Bodoll mm LLC**
Book: Hard copy, E-book; global distribution
Kindle, Nook, IPAD, & Digital Publishing formats.

The Book is about the water festivals & sporting event of boat races held annually on the Kuttanadan rivers and canals heralding the joyous arrival of the folks' festival season of Onam. The book narrates the beauty of the land, long history, traditions, nuances, community participation, pride, fervor, mood, the rhythms, attractions, and the compelling nature of competitive spirit always inherent in them.

AVAILABLE IN 3 DIFFERENT FORMATS

1.Print Book **ISBN:978-0-578-25679-5** [Book Trim size: **8.5" *11" hard copy**]
Pages 400
Picture collages (36)

2. E-Book **ISBN: 978-108-799-0613** Format:-**EPUB**
In all popular ***Ebook*** formats

3. Download Web site: ***https://www.desi-paradesi-diaspora.net/***

Distribution & Availability: **Global** distribution by IngramSpark retail network
Digital download in compliance with "DMCA" guidelines
Copyright information: protected under the copyright laws
Preface/Synopsis online, free reading
Chapters described under "Synopsis"
Glossary English
Native words

Index	provided
Barcode	
Disclaimer	listed on website
Terms & conditions	Listed on website
Privacy Policy	listed on website
Web site	***https://www.desi-paradesi-diaspora.net/***

Coming Soon!

- ***Singing simplified!***

Tips to improve singing abilities with the actual resources each of us possess

"Each sound is unique and is music to one's own ears!"

- ***Namasthe, India! Let's play the Amazing Accordion!***

Book offering text and video instructions for playing the instrument

Vincent, the magnificent

A story of a boy from the fishermen's community on the Arabian Sea cost, who has grown more practical and accomplished than his peers, but became largely isolated by lack of recognition. *Please check back for new additions!*

The multimedia website include literary works of:
books, anthologies, essays, articles, stories, poems
music works of lyrics, songs, parody music of internationally well-known songs, music score sheets, music videos, music translations,
***Video tutorials,**
presentations, mp3 audio files, and graphics

Contributing YouTube channels:

https://www.youtube.com/results?search_query=bobbyalis

https://www.youtube.com/results?search_query=bodoll+mm+llc

https://www.youtube.com/results?search_query=bobbyalis

https://www.youtube.com/channel/UCHaTFGLY184cLN9y4U2d57w

Other resources on the Website: *https://www.desi-paradesi-diaspora.net/*

Several multimedia items and other interesting literary works such as poetry, anthologies, short stories, lyrics, songs, along with contextual summary translation of several internationally well known songs are available.

A number of **music sheets, music videos** , accordion video tutorials, etc. are posted on the website.

Potential Sponsors, and advertisers:

Television, Movie, Music, Science, Engineering, educational, environmental energy, technology institutions, NRI service Banks, hospitality, International Airlines, Travel & Tourism, publishing

Alphabetical Index

Vernacular words

The Book is Sponsored By:

The Pallithanam Heritage Bungalow, Homestay

Author bio

I'm now retired; feel blessed to be able to live moderately active. Prior to that, I was an engineer in Electronics & Telecommunications, and had worked in various Television fields. I had the opportunity to work in 4 countries in 3 continents which necessitated me to learn to speak 4 languages. The highlight of my career and life was when I worked in the ME for a tenure of 17 consecutive years among an International, close-knit residential community of scientific, medical, research professionals, and their families. I had the privileged opportunity to work in advanced state-of-the art technologies in professional work environments receiving recognition. My work & study in Germany among great many numbers of multinational 'Gast Arbiter' workforce and native Germans perhaps paved the way for my selection. Both have been stimulating, rewarding, and beneficial experiences. I decided to share the wealth of knowledge amassed, and the experience gained during these years, before the memories fade away and the time runs out. I hope that a few readers could be inspired and enriched by them!

There is no better way than telling them to the global community in stories written in eyewitness authenticity, feeling *worldwide* vibes, pulse, perspective, sensitivity, and sensibility. An ancient oriental maxim states: "Sharing knowledge and experiences, are aggrandizing and enriching; benefits others. It's a convenient path to attain immortality!" Currently, I'm preoccupied with writing and trying to do more of it and better.

Sensing some dormant literary talents, I extended my foray into essays, short stories, poems, parody lyrics for internationally well recognized songs; scoring & recording music videos of songs and genre of different cultures for the enjoyment of the international communities. They are posted on my website. For these activities, I chose "*Aurora Bobbyalis*" as my pseudonym, while my immediate relatives and close former friends can still recognize me by the original, uncorrupted, baptismal name!

I have a habit of getting interested in several things at once, some of them have become obsessive and even distracting from the others currently working on. Some of them might have got stuck in the mind while pursuing career and making a living. Those times, one doesn't usually find time or energy to spare and pursue new frontiers.

Though most of the subjects are somewhat related, they require special talents and skills. I'm now trying to satisfy my longings by developing them into enjoyable and active hobbies, though at times, find them quite challenging.

The publishing of this second book is a fitting testament to that cause. I was able to assimilate several of my career experiences in Television, Studio technics with the music related hobbies. As a result, it provides the readers a detailed, realistic account of the multi-faceted technology involved in sound enhancement, music appreciation, scientific aspects, Television & Sound production technics, and nuances in stage performance. Besides, the book reads like a novel, narrating several sequences in graphic events, expressing true emotions, and synergistic interactions bringing out drama, suspense, happiness, wonder, and entertainment. Each episode presents a certain moral lesson, a piece of information; highlights the genuine human response or outlook. Exciting reader reviews are received from across the demographic spectrum, especially from professionals engaged in Fine & performance arts; TV/ Film & Sound recording. Many stated that they enjoyed the extempore adjunct-events and the drama in the book that continued on, even after the end of the

magnificent concert. An enigmatic and investigative curiosity still prevails among them to fathom the mysteries created. The organizers were excited about the exuberant responses from the audience throughout the concert. The Millennial teens loved the tormenting and nostalgic story recited by Elise, besides her fabulous singing. The elders, in particular, appreciated the introduction of the theme stories and the setting of mood to the upcoming music piece. It evoked strong eagerness to listen to the songs in the next immediate proximate moment!

The audience relished the grand concert; the supreme sound qualities within the walls of the hall; the fabulous internal & external surroundings; to crown it all, the zestful immersive audiovisual experience, and the emotional involvements they had.

www.ingramcontent.com/pod-product-compliance
Lightning Source LLC
LaVergne TN
LVHW081945200926
848484LV00009B/2510